ROLAND WOLDT

Successful Process Mining Projects

A practical guide on how to do good process mining analysis

First published by What's Your Baseline? 2025

First edition

ISBN: 978-1-967221-04-2

This book was professionally typeset on Reedsy.
Find out more at reedsy.com

For Tanja, Paul, and Carl

By fire. That’s the best way, really.

— Ethan Hunt

Contents

Foreword

If you have ever worked in or come into contact with the incredibly interesting topic of change management, you will have heard about a certain principle called single-, double-, and triple-loop learning. Single-loop learning pretty much means that you correct whatever you did wrong, and then later on you make the same mistake again, without learning from it.

Henry Ford said it a long time ago: "If you do what you always do, you will get what you always got".

Double-loop learning enables you to figure out what the root cause is of what went wrong in the first place. If this mistake happens time and time again, it might be more efficient to take away the root cause instead of correcting the same mistake over and over.

With triple-loop learning, you'll go one level deeper, and you'll try to find the underlying cause for the root cause that you discovered in double-loop learning.

Sounds a bit theoretical? I agree, let's make it a bit more tangible.

You are responsible for the performance of a business process, and let's keep it simple: the procure-to-pay process. Now imagine that you keep having problems with matching the invoices to the purchase orders, and mind you that in sizable organizations we could be talking about hundreds of thousands of invoices per year. When applying single-loop learning, the people who are actively engaged in your process might be executing the same corrections to the invoices time and time again, burning a lot of time in the process and thus costs. So, you take inventory of why things require a correction and find out that one of the root causes is the fact that there frequently are no purchase orders to match the invoices to. This is already a nice example of double-loop learning, and the corrective action to avoid this in the future is to send the invoices back and notify the supplier that they are required to receive a purchase order from you if they intend to get paid.

Ultimately, if you apply triple-loop learning, you might come to the conclusion that you have failed to put decent contracts in place with your supplier in which it is stipulated that the use of a purchase order is mandatory. So, you take away the underlying cause of the root cause of your symptoms.

Now, what on earth does this have to do with process mining?

Well, more than you think. The mere act of finding out what your organization actually has been doing gives you a tremendous amount of insight into the symptoms of process sub-performance. If your data set is rich enough (in other words, it contains sufficient related data points), process mining can also provide you with valuable insights into the root causes of these symptoms. Finally, if you have the luxury of running process intelligence operations (the combination of process mining and process management), you also have the business context at your disposal, and you can find the underlying cause for some of your symptoms.

So, where does this wonderful book come into play?

Setting up a process mining operation is not the easiest thing to do. Granted, it's not rocket science either, but the data collection, ingestion, transformation, analysis, and interpretation can be daunting given the enormous amount of data available—and I am not even going down the rabbit hole of the organizational aspects of implementing and running process mining. Roland's book gives a well-structured and easy-to-follow approach for understanding the basics of how to get your process mining endeavor launched and in orbit. If you find yourself at the start of this amazing journey, this book is your tour guide!

Enjoy the read, and thank you, Roland, for giving me the wonderful gift of providing a foreword for your second bestseller. Keep on writing!

Caspar Jans

Principal BPM Expert and Global Head of Celonis Process Management GTM

Co-host of the BPM360 podcast

Ex-colleague of Roland

Preface

Let me tell you a story: the first time I, as an experienced German Autobahn driver, came to the United States, we rented a car in San Francisco at the airport and drove South on the interstate to our hotel. While doing that, I wondered why everyone was moving so slowly because my speedometer only registered 80 kph on this five-lane road (the legal speed limit for trucks is 80 kph, and 130 kph is the recommended speed on the Autobahn).

Well, I realized that these 80 were not kilometers but miles per hour, and I could be in trouble after a few minutes behind the wheel. Luckily, no blue lights turned on behind me, and I completed the trip to the hotel, swimming with the flow on the road and arriving safely.

Why do I tell you this story? It is an example of how a small technical device in your vehicle that measures data shapes your perception and behavior and has a significant impact on the output and quality of the process that I was executing—I drove safely and within the expected time to my destination and accomplished the goal of that process.

But where are we in this regard in our organizations?

- Do we measure the performance of our processes? Well, often I see some metrics being measured (like "number of widgets created" in manufacturing), but those are mostly lagging metrics.
- There are almost no predictive measures being derived or defined, and processes are not adapted to changing situations, for example, when it comes to delays of a supplier shipment.
- And then there are just a few organizations that actually have their processes formally designed, for example, in the form of process models. And those who do might see them only as an exercise to "appease the

gods"—may that be the management or the regulators.
- That leads to interesting results when it comes to audits—some poor souls trying to figure out what actually happens in the processes that they audit. And that might lead to regulatory notes that your organization has not made enough progress or even to hefty fines (I am looking at you, finance industry).

Overall, I am missing an awareness of the need to measure what you are doing. Some smart person once said, "You can only measure what you see," and how boring would a football match be if we had not defined what "winning" means and how you can create the necessary points that will end up on the scoreboard? But why do so many organizations not apply this to what they are doing?

Process Mining to the rescue

Over the last couple of years, I have seen more interest in answering this question. And for good reasons:

- Process mining (as the main technology for process intelligence) is the "speedometer for your business."
- It will give you the visibility into how your processes are executed. Not only on the "happy path" that you might or might not have documented, but with all its variations and loops that you have stored in your runtime systems of record.
- Process mining gives you calculated performance metrics out-of-the-box, like frequencies and times, and also shows you where you have rework or other unwanted behaviors in your process data.
- Moreover, you can create your own dashboards for calculated metrics that allow you to investigate your hypotheses of your process performance. Data from process mining analysis can also be sent to traditional business intelligence tools and be embedded in already existing dashboards as well.

And moreover, you can trigger actions from thresholds that you define in your process mining tool and interact with running process instances—think notifications, alerts, or transactions.

This also allows "non-process nerds" to gain insights into what they are doing in their processes.

- Lastly, process mining is also embedded into a larger Business Process Management (BPM) practice and tool stack, and you can, for example, download discovered processes (in case you did not have any documentation) and use these for future-state process design and simulation. Or you can upload already existing reference processes into Process Mining from your repository, and Process Mining checks to what degree you have conformed to "how the work should be done."

Mining technologies close the missing gap in the process/solution lifecycle and provide the data-driven analysis approach for your complete process performance management.

Who is this book for?

This book is for people who are serious about how they can improve how their organizations run and how they can make their next large transformation project a success and be able to measure the benefits of the large efforts that go into that.

It is for the process and architecture practitioners who run their programs for years and don't get the visibility that they deserve.

It is for the analysts who want to switch their practices to a more data-driven approach (while not forgetting the human contributions to any analysis) and for whom the words "Digital Gemba" sound good.[1]

And of course, it is for the curious folks who have heard about "process intelligence" and were wondering what this means and how this approach and new technologies fit into their organizations, without creating another hype and disappointment when the results of the first project do not meet the inflated expectations ... as you know, the first time you do something, you most likely suck, and that is the same here.

But wherever you are on this map, this book has something for you and will help you decide when (not if) and how you will start implementing mining technologies. The book covers these areas:

- What is process intelligence, process mining, and all those fancy terms that you might not have heard about? And more importantly, it helps you decide what, if anything, is needed to get started.
- A six-step approach to process mining (as the core technology of process intelligence) that will help you to create value from your first project on.
- An introduction to the methodology and tools that you need to do data-driven process analysis that is more than just "throw some data into the tool and let's see what happens"—an approach that definitely will lead to failure ;-)
- How to select, buy, and implement mining in your organization. Who is involved? What can you expect, how long will it take, and what effort is needed? And so on.
- A hands-on walkthrough of the steps that you will encounter in a project. Even more, you can download example files, and I will point you to basically free-of-charge tools that you can use to recreate the scenario in this book at your pace.
- This will also be supported by freebies that you can download from the book's web page.

But why should you trust me?

That is an excellent question, and given the fact that within a few weeks all LinkedIn experts switch from one topic to another, there is some reason for concern. Remember how you don't read about the blockchain there anymore, but everyone praises the world-saving technology of AI, even though it changes so fast these days that we don't know what it actually does? Process mining is (was?) one of these hot technologies, too. And all the vendors now combine AI into it as well... Oh, my ...

Well, I have been working in architecture/process management for almost 30 years now with various large organizations whose names look good on a resume—half of the time with two vendors in that space (IDS Scheer/Software AG/ARIS and iGrafx), and the other half with big consulting firms, Accenture and KPMG. And while doing this, I worked on about 60 projects with brand-name clients,[2] helping them to be more successful and apply the tools that were state-of-the-art in their organizations.

One of those tools was process mining, and I was first exposed to it in 2004 when we did a performance management project with the German Armed Forces that should help them in their transformation and establish Service Level Agreements (SLAs) to determine the right size of employees and soldiers based on the processes that we designed and implemented in SAP. And we were using the first commercial process mining software for this—ARIS Process Performance Manager, which unfortunately was too early on the market to become a success.[3]

Fast-forward many years, and after many BPM projects and helping in sales, seeing firsthand where clients' heads were (and how they reacted to vendor messages), I became the Head of Software AG's Global Process Mining Center of Excellence. And then I moved on to iGrafx who had just bought LogPickr, and we integrated that tool into the Process360 Live platform.

So, long story short, I have worked with process and task mining tools for many years and learned that the first steps into this new technology might be the make-or-break point for an organization. Get it right, and everyone wants to be the next in line. Get it wrong, mining technologies are put away as the next fad, and the way "how we have always done things over here" takes over (again).

That experience triggered me to write this book because, unfortunately, I see too many clients still being on the fence, not knowing how to get started, and thinking that it takes millions of dollars to get off the ground. Nothing could be further away from the truth, as you will learn in this book.

So sit back, buckle up, relax, and enjoy the ride!

Acknowledgments

Practice makes perfect, they say.

That is true to a certain degree, and it was easier to write this book after I had already been through the whole book writing, editing, and publishing process. However, writing and publishing a book is also a team sport, and I am happy that I have the best players on my team.

These fine people make you a better author (and person, FWIW) by helping you to identify potential topics and shaping your thinking about those. They also challenge your assumptions and reasons why you want to do this in the first place. And they also graciously share their experiences when it comes to business aspects, as well as giving you moral support when things don't go as fast or as well as you expect.

I am glad that I have these fine folks on my side:

- **J-M Erlendson**, who, to nobody's surprise, is my main supporter for all things What's Your Baseline. He and I have run the podcast for well over 100 recordings and over four years by now, and every single interaction with him is funny and insightful.
  He also graciously read the audiobook version of my first book, "Successful Architecture Implementation," and it became a better version because of this (and it also doesn't have a German accent that you would have to endure for hours on end). ;-)
- **Caspar Jans**, the author of the foreword. He and I worked together at Software AG a few years ago, and we both have moved on since then but stayed in contact. And I take a bit of credit for him starting his own BPM360 Podcast—go look it up in your podcast app of choice. You will like what he and his co-host Russell do there.[4]

- **Martin Holling, Katrin Wetzel, Randolph Schütte, and Marcus Zschuckelt** who are the reviewers of the German version of the first book. Together, we learned a lot about (AI-supported) translations, technical terms, and that you cannot simply translate things one-to-one. If this book might get a German translation, I know whom to ask for support.
- The **beta readers of this book**: Anthony Phillips, Chris Kim, Hanneke Loefs-Mos, Katrin Wetzel, Khemindra Nadarajah, Marc Mendelson, Marcus Zschuckelt, Martin Holling, Matúš Mala, Michael DeCamp, Nick Blackbourn, Per Bandsholm, Randolph Schütte, Ulf Henningsen, and Zbigniew Misiak. Thanks for your review of the manuscript and your valuable feedback—this book is better because of you.
- All **guests of the What's Your Baseline? podcast**.[5] Your ideas and gracious donation of time allowed me to learn new things, question my assumptions, and think about things through your eyes. And moreover, we're having a lot of fun recording our shenanigans.

And lastly: **YOU**—the reader of the books, the listener to the podcast, the follower on LinkedIn and elsewhere. I would have never thought that that small idea of me writing articles and trying out a different form of publication would lead to over 50,000 downloads, two books, speaking engagements, and the creation of a formal company, the What's Your Baseline LLC.

I love to connect with you and enjoy when someone reaches out to me out of the blue, either to discuss an EA/BPM/mining topic or simply hang out with each other and chat about whatever is on our minds on that day. Keep it coming!

But don't stop at that. Connect to each other; meet in person or online to exchange ideas and make our discipline a better place. We all share this goal, despite where we live or work—we are all humans, and human connection is what counts.

I

Process Intelligence Introduction

1

Process Intelligence Introduction

We are living in interesting times. On the one hand, we see everyone jumping on the AI train and following the promise of a fully automated future. On the other hand, we run our businesses like "we have always done"... not knowing what really happens and making decisions based on reports of what has already happened.

This is a rearview mirror perspective, which feels so antiquated these days, while everyone in high-tech is talking about data, and we are consuming data all day. Why don't businesses do this?

"Process Intelligence" is the latest term that is pushed by the vendors in the process space and the analysts, and it has a quality that everyone can read everything into this term. However, when you take a closer look, this narrows down to the dominant new technologies in this space: process and task mining.

What's in this section?

This section sets the stage for the book and will clarify not only the different definitions but also explain how process mining fits into a bigger picture of process management and improvement. Things that never go out of style in well-managed organizations.

We are looking at:

- What is process intelligence, and why should you care?
- Process mining in the larger solution/process lifecycle
- The six-step approach to successful process mining projects

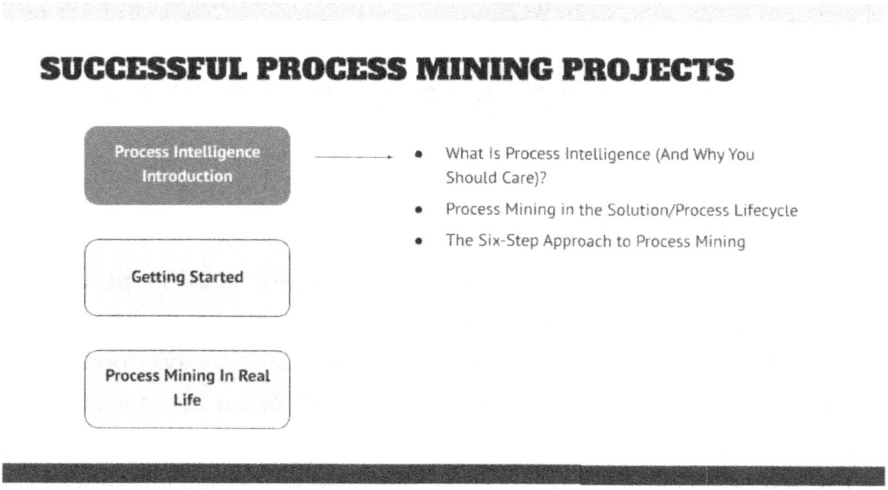

Fig. 1.1: Process intelligence introduction structure

2

What Is Process Intelligence (And Why You Should Care)?

Process intelligence is a term that you might have encountered a bit more often these days, and one big driver of this is the popularity of process and task mining. However, there are also voices out there who think that there is more to process intelligence than mining and some who think it is just a subset of traditional business intelligence (BI) and wonder what all the buzz is about.

This chapter attempts to bring some clarity into this conversation, starting with an analyst definition and a spin on this. While then looking at the underlying problem, the new(-ish) approach to it, and the different methods that have existed in some cases for decades and already had their "day in the sun." This will give the context for the main topic of this book.

What is process intelligence?

Let's start our conversation by looking at how Bernhard Schaffrik from Forrester defines process intelligence:

> *"Products that provide analytics about an organization's processes that are operated by humans as well as implemented in IT systems. Process*

> *analytics is mainly based on task- and process-mining capabilities but not limited to them. Process intelligence software enables decision-making on how to improve process performance.“*[6]

Interesting enough is that Forrester is focusing on the software more than the discipline. Which might be a function of the context in which the analyst report was written, but I have seen many instances where clients saw process intelligence as a pure software category spiel. However, nothing could be further from the truth, and we will discuss the different aspects of process intelligence further below. In the following chapters, we take a look at the six steps for a successful process mining project.

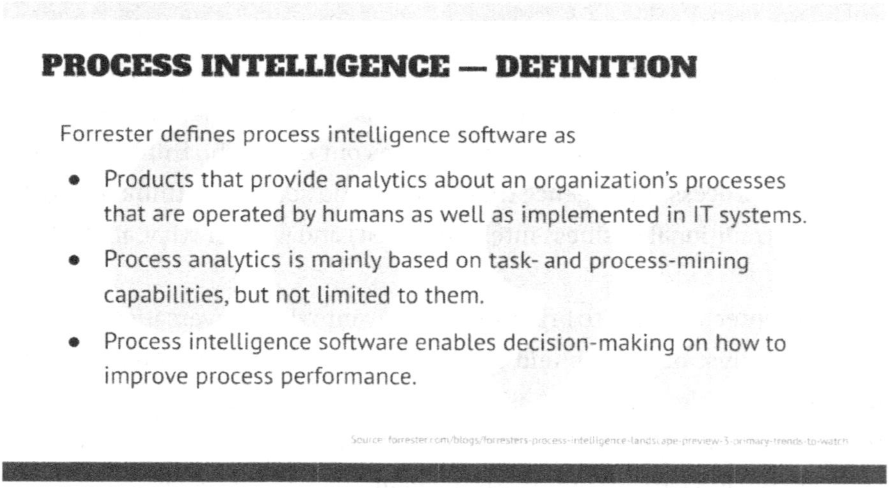

Fig. 2.1: Forrester's definition of process intelligence

So, what is the specialty of process intelligence according to Forrester? It is the mining technology, the focus on the complete process that includes not only system tasks (which might be a challenge to capture if you just use process mining software and not also task mining software), and the focus on decision-making ... but notice the word “enabling” in the third bullet—he didn't say "provide." Which is a reflection of the fact that there is a lot of human work to be done to get to the point where one can decide about process or other improvements that then later will be implemented in projects.

What is the problem?

Let's take a step back and take a look at the difficulties that exist in how organizations run their processes.

There is a considerable **lack of visibility** in all processes that don't create physical goods that you can watch, like a production process or a retail sales process where you can see the capacity of the process with the naked eye, answering the question of "What is the output of my process?" For most processes, the output is partially hidden—it is the number of loans that have been underwritten or the number of candidates being hired and ready for onboarding. We also cannot easily see how long a process or its instances take and if service level agreements (SLAs) are met.

But these are not new issues, and there were attempts to bring in this visibility by using **process analysis** techniques and BI in the past. And many people have made money by providing services to create visibility and analyzing processes in the past.

The ultimate goal in all of these efforts was (and is) that you **change the execution** of the runtime processes and create results and changes. This can be changes in the process itself, the organization, systems, etc. And you need to be able to measure the performance of these changes then as well. The graphic below shows the issue in an overview.

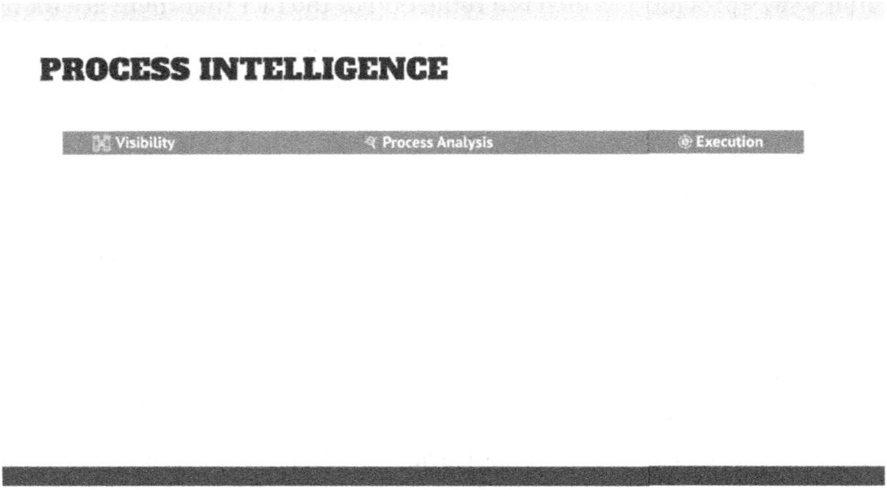

Fig. 2.2: Process intelligence problem

The new characteristic of process analysis

The key aspect of the issue with process visibility is not that we cannot measure process performance—back in the day, people were observing the process with stopwatches and gathering the results on sheets of paper. There is a new focus on technology that allows an automated capturing of visibility data that has a much higher quality and objectivity than anything else that has been captured manually before.

Mining technologies allow capturing **all** activities as Forrester has defined above, independent of whether they leave traces in systems (process mining) or are captured by observing unstructured activities on user machines (task mining). They also calculate metrics that can give context and insights out-of-the-box, like duration (minimum, maximum, average, median times), and so on, which would take some effort to calculate otherwise.

This allows analysts not only to get to results faster, but they also are based on facts that are in the data set. There is just a limited amount of conversation possible—for example, when the context of a situation is needed when you see an interesting visualization in the process mining charts. This improves the quality of the decisions made significantly because the subjectivity—"Is what I hear from that subject matter expert the main information, or is he telling me what he wants me to hear?"—is taken out of it.

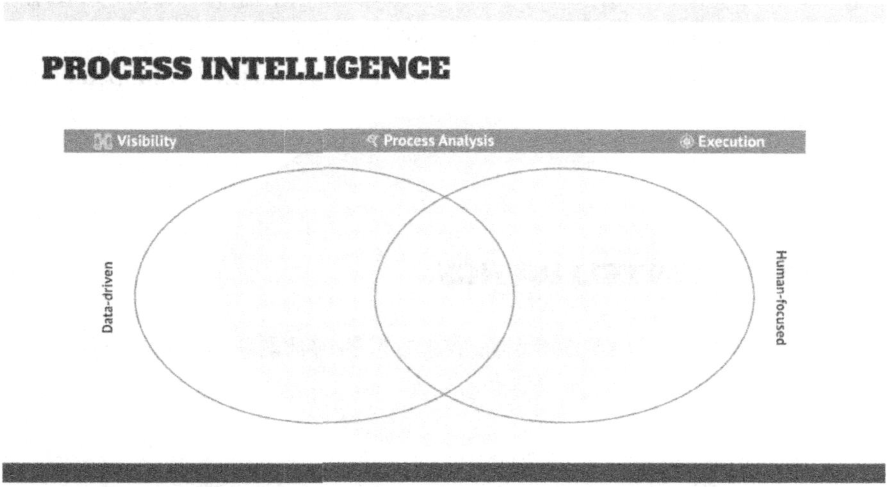

Fig. 2.3: Two poles of process analysis

The big challenge with mining (and other) tools is that they just *support* decision-making but don't make the decision for you. You could trust generative AI tools for this, and we will see a lot of them being built into any tool that we will use soon. But at the core, they are just guessing the next word of an answer by calculating probabilities.

Would you risk the future of your company by trusting a probability calculation that sometimes can lead to "hallucinations" of the AI software? I am skeptical of this and would rather see AI support more traditional process analysis techniques as "little helpers" that save time in doing the busywork but have a human identify recommendations and make decisions on changes.

How do various analysis techniques fit into this continuum?

Process analysis is not a new thing, and some methods have been around for decades and had their heyday in some cases decades ago but are still relevant. And we will see an integration of AI into the underlying tools, which might make them more powerful and accessible for users.

The figure below shows various process analysis methods mapped on the Venn diagram:

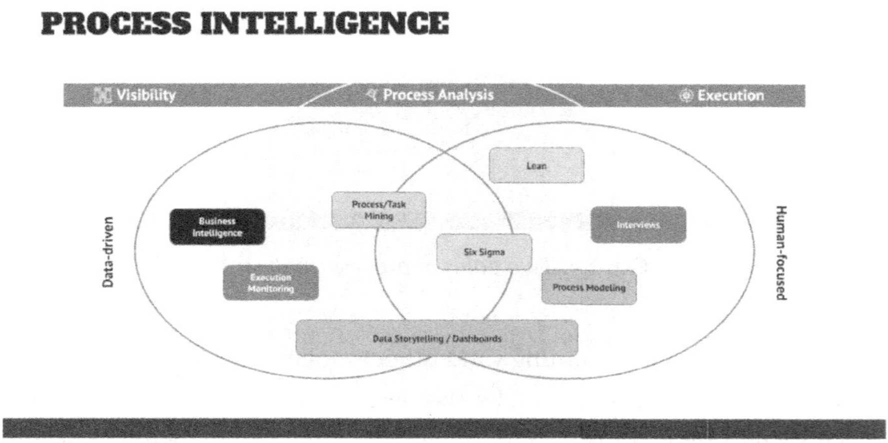

Fig. 2.4: Process analysis methods on the continuum

Let's take a brief look at each method:

- **Business intelligence** is the practice of creating visibility in how the business is doing by connecting to source systems and visualizing metrics in dashboards. It is the grandfather of what is done now in modern mining tools, but the success of it depends on the creativity and skill of the creator of a dashboard to convey the right information to the right stakeholder.

One additional drawback of BI is also the fact that mostly lagging indicators are used for the visualization, and predictive technology is not widely implemented in the BI tools. They are also not necessarily process-oriented, even though some vendors provide process mining plugins for BI tools like PowerBI or Qlik.

- **Execution monitoring** is typically built into automation/workflow tools and can be configured to show process metrics. Unfortunately, they typically are focused on the currently running processes in the system (that run in the one system that has this functionality built in but might have difficulties including data from other runtime systems in the process) and are not focusing on creating a higher-level summary of all processes and helping to identify patterns of process performance.
  Independent of the shortcomings, they provide insights into the execution and have alerting/notification capabilities if something goes wrong. And some runtime tools also allow exporting data that then can be used in traditional BI tools or dashboards.
- **Process/task mining** tools fill that gap. They are designed to be process-centric and are fed by multiple systems in the process. However, they might have different levels of analysis tools built in and cannot bring in non-process metrics for visualization purposes.
  Some process mining tools can export data to BI tools and have machine-learning/AI capabilities that can give predictions on how cases might end in the future. They can also trigger notifications/alerts and trigger actions in runtime tools. However, they are not necessarily complete regarding their analysis capabilities, and simulation or detailed statistical analysis might be in a different module of a vendor's platform.
- **Six Sigma** is a method to fill this gap. It is meant to calculate deviations in metrics that typically are used to improve the quality of processes ("How well is it executed?") and reduce defects.
  Sporadically, this can be a bit "nerdy" for the regular analyst and go into statistical details that they might see as "overdoing things." Independent of that, the statistical and data-driven approach that Six Sigma brings to the table will improve the quality of the analysis, and a

solid understanding of the statistical approach will be helpful in mining projects as well.

- **Lean** is another methodology that aims to find waste in processes, such as overproduction, defects, inventory, etc. Those shall be optimized and can be helpful to give your mining projects an initial focus if your client has trouble finding hypotheses for (bad) process performance.
  It also comes with various visualization and analysis approaches, like Fishbone diagrams, that are also helpful in other analysis types. And it is often combined with Six Sigma. I highly recommend gaining a good understanding of the Lean method for the average business analyst without the need for a formal "belt" certification (Green Belt, Black Belt, and so on).
- For a long time, **process modeling** was the analysis technique of choice—visualize the content in a diagram and find consensus on how work is done or should be done. Add objectives and measurements to them (current state and future state), and run a couple of reports on the diagrams to find improvement ideas.
  In a more advanced approach to this, tools like **process simulation** were used to create a performance baseline and create scenarios to resolve bottlenecks and other unwanted behaviors of the process.
- **Interviews** are the easiest analysis technique—talk with as many people as possible (process performers, process owners, system owners, etc.) and see what "bubbles up" or if your initial understanding of the problems is validated.
  Even though it sounds as if interviews are the least professional analysis technique, they are still needed in many other approaches as well. The interview partners who have done the process for many, many years will give you the context and the details of the process and thereby deliver the context for your analysis and ultimately the improvement opportunities that you will identify.
  However, the risk with this technique is that you are dealing with people, and they are highly subjective. They might just tell you what they want you to take away. The majority of process instances might run smoothly

within the defined service levels.

- The last technique on the list is relatively new and might be applied to all other methods: **data storytelling and dashboarding**. While dashboards are used in BI and other approaches, they might be considered a technical thing that visualizes information. The risk is that there is no context to the data that is shown.
  The context that you get from interviews will allow you to tell a story with the underlying data and therefore transform it into information and knowledge ... and sometimes wisdom when you can see what you might not want to do ;-)
  Data storytelling is a skill that I have not seen in many business analysts or dashboard developers, who might approach the creation of them from a more technical view ("Which widget do you want for that data, and where on the canvas should it be?").

As you might have guessed by now, to do a proper process analysis, you will have to take pieces of each methodology and combine them to fit your situation. If you are doing a process mining project, for example, you will have to do interviews with stakeholders but also extract and transform process data to show them in a dashboard. These are aligned to the hypotheses of process performance that drive your analysis.

What does that mean for implementing process intelligence?

Given the information above, it is clear that the Forrester definition falls a bit short—and it slaps the title "Process Intelligence" on process/task mining tools. When you start with your process intelligence implementation, I recommend focusing on the three steps that we outlined in the beginning:

- **Create visibility**: this can be the creation of dashboards (as the BI folks do) or the creation of models and diagrams, which is what traditional business analysts have done, and then add performance data to these artifacts.

In both cases, the objective is clear—get a performance baseline and make it visible because only what you see can be improved.

- **Analyze the process**: combining data-driven and human-focused approaches is the right way going forward. What you want, though, is an analysis result that is aligned to objective goals and not being tainted by the perspective of "important" stakeholders. Your recommendations of opportunities should be based on facts and be reproducible and understandable by another person.
- **Execution**: The best ideas are worth nothing if you don't change things and improve them. This means that your recommendations need to be realistic from an implementation perspective—this includes things like feasibility, cost, or organizational/technical readiness for your ideas.

  When you implement your change, you should always think of the key performance indicators (KPIs) and metrics that you have defined. Make it an integral part of your solution design to capture those and compare them with the baseline metrics that you identified in the previous steps and were used in your process analysis.

And what are the main use cases?

But what does that abstract definition stuff all mean? What I've seen in process mining projects are the use cases that are shown in figure 2.5 below (and this list does not attempt to be complete; I am sure that there are more use cases out there):

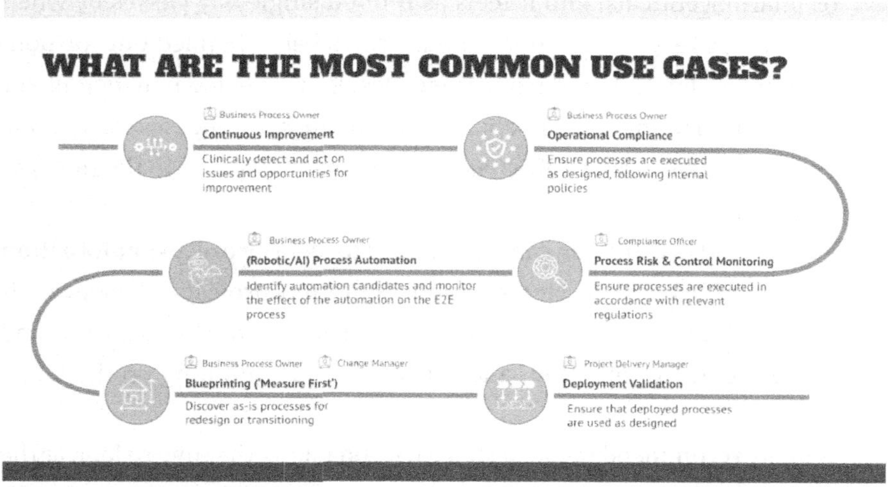

Fig. 2.5: Main use cases of process mining

- **Continuous improvement** is most likely the most popular use case that drew you to process mining. The data-driven analysis that will add more quality and objectivity to your results and will bring new tools to the tool stack that we've discussed above.
- Another use case is **operational compliance** and thereby maintaining a consistent output and quality of the products and services that you provide. Customers expect consistency, and we all have seen how changes in the supply chain and trying to "make the best with what we have" during the pandemic interrupted this and led to recalls and other after-the-fact measures that cost a lot for the manufacturers.
- Related to the operational compliance, which has the process owner as the main stakeholder, is the **process risk & control monitoring** that is the focus of the compliance team in your organization. It ensures that processes are followed as designed while observing all compliance requirements that are defined by the regulators and manifested in your policies.

  This is especially important in heavily regulated industries like finance or pharmaceuticals, and it feels as if not a single day passes by when you do not read about billion-dollar fines being handed out for non-compliance by the big organizations. Besides that being non-compliant hurts the customers, it is a shame that so much money is wasted if one can implement a relatively cheap technology (compared to the fines) to avoid that situation.

- Process and task mining can also be used to identify **process automation** opportunities and accelerate implementation projects. However, it appears that these days we forget to do this when it comes to AI and other robotic automation, and everyone is just jumping on the latest tools.
  I highly recommend taking a step back and taking the time to look at the process to decide what you want to automate. The fact that you can do this with AI does not mean that you must do it—I rather recommend comparing the various automation options and seeing what the impact of these will be. Simulation, which we will talk more about later in this book, is a good tool to make these predictions based on the numbers that you get from process mining.
- This fits nicely in the overall **principle of "measuring first" and creating blueprints** for any implementation project. Creating a business and technical blueprint before you start "wrenching" (or, to say it more professionally, creating a solution architecture) took a backseat in organizations over the last 10-15 years. Bring it back and start with measuring the process performance as a starting point ... we will discuss the "levers of a process" a bit later when we discuss the six-step process for successful process mining projects.
- Lastly, you can use process mining for **deployment validation**. I have seen many projects that dissolved the project team a few weeks after go-live and the immediate rollout support. But how do you know if you realize the benefits that were promised in the project's business case? How do you know that the users follow the new process, supported by the changes to the systems that they use? I guess you already have an idea—process mining to the rescue :-)

3

Process Mining in the Solution/Process Lifecycle

Process mining does not exist in a vacuum. It is part of a larger lifecycle of how organizations manage their processes. Some organizations do this in a structured way, and some do it "invisibly" (read: don't manage their processes at all).

This leads to the situation that process mining is considered a technical exercise and not as a part of something "bigger." This chapter will give a brief introduction to the solution/process lifecycle that is comprised of five distinct phases.

Two areas—design and execution

The first thing that one needs when looking at the solution/process lifecycle is to distinguish the fact that there are different tools for different purposes, but in general, they fall into two categories:

- **Design-time tools**: These tools allow you to create blueprints or analyze existing content. The artifacts in this area will also be used to compare the long-term execution with the initial objectives. Classic architecture tools, such as ARIS or HOPEX, as well as the mining tools, fall into this category.

- **Runtime tools**: These are the actual transactional systems and support systems that allow the monitoring of them. SAP, Salesforce, Dynamics, Workday, and other systems fall into this category. Depending on the capabilities of the actual runtime systems, you might have to complement them with BI tools for monitoring.

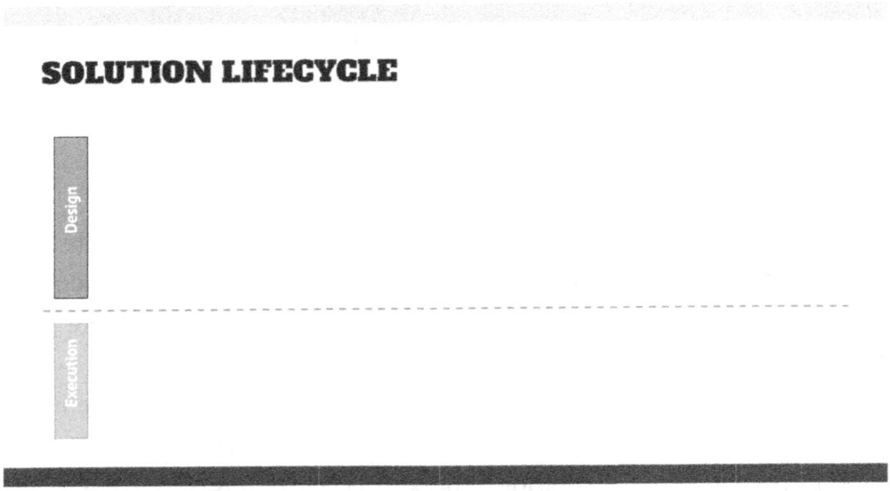

Fig. 3.1: Two areas for processes

The one mistake that I see clients make is to value one side over the other. Of course, you cannot run a transaction from a modeling tool. But without a proper design that can be critiqued by others (so that it shows the implementation design and plan in a granularity that there are no misunderstandings when it comes to configuration, coding, and testing), the implementation is already falling behind. On the other hand, many architecture-focused teams tend to be too detailed and strict about what needs to be created for approval. The implementation gets delayed unnecessarily, and people perceive architects as a roadblock instead of a valued contributor.

Phase 1: Strategy and the Enterprise Baseline

In a perfect world, one would start with a clean slate and start creating the repository of what is currently in production in all architecture views as defined by your chosen framework (such as TOGAF).

In this phase the structure will be set up, the data and meta models being defined as part of the technical governance phase of the implementation, and existing content will be imported and normalized.

The main idea here is to create the "Enterprise Baseline" as the always current version of "what runs."

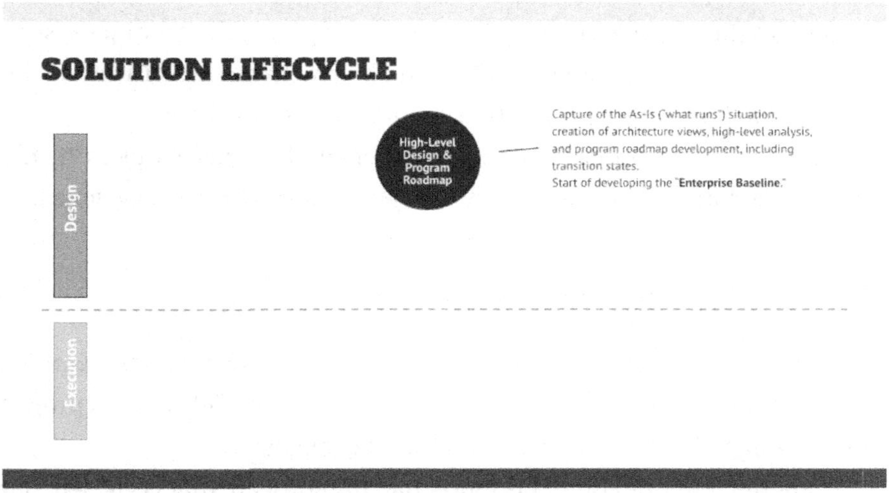

Fig. 3.2: Strategy and enterprise baseline phase

The good news is that you don't have to create everything at once, which would be a Herculean task that would also reinforce the impression of the "ivory tower architects."[7] The way to avoid this scenario is to provide "just enough architecture" guidance (principles and a minimum of reference architecture content, such as preferred technology). Work with your current projects on the definition of a minimum architecture structure (formal and content-wise). Bring in their designs into the repository—in the "Projects/Solutions" area, and in the case of approved designs or projects that are close to go-live/just went live, into the "Enterprise Baseline" area.

You can also use modern technologies like process mining and task mining (robotics process discovery) to identify current state content and create straw man models to complete an initial picture of the architecture. You should also think about importing the current strategy artifacts so that you have a baseline set of objectives, requirements, needs, KPIs, etc., that you can map your (future) projects against.

The main purpose of the enterprise baseline is to provide the foundation for the identification of change and improvement needs, for example, to reduce technical debt, to enable process improvements, or to define the structure changes of the organization. In addition to this, analysis techniques, such as process simulation, can be applied and help with the identification and justification for a future project (to support the business case).

These identified projects then will be created as placeholders in the "Projects/Solutions" area until they are approved and further developed.

Phase 2: Solution architecture and design

In this phase of the solution/process lifecycle, the solution that was identified as a potential problem solver in the previous phase and defined as a project will be designed. The result describes "what changes".

A "solution" in this context is more than the software that is chosen, but it is the summary of all views (processes, apps, data, org., etc.) needed to implement, update, or retire one or more capabilities. Therefore, it is crucial that a project include the strategy artifacts needed and not directly jump into business and technology blueprinting. Having clarity about objectives, KPIs, business models, etc. will help in coordinating and resolving conflicts with other programs and projects that might have a less well-defined scope and program approach.

It describes the design for one transition state, and it is recommended to tailor your projects in a program accordingly by transition states related capability configurations. This allows controlled change management going forward while coordinating your program's schedule with the other related programs so that the continuous operation of the implemented processes and systems for the end users can be ensured.

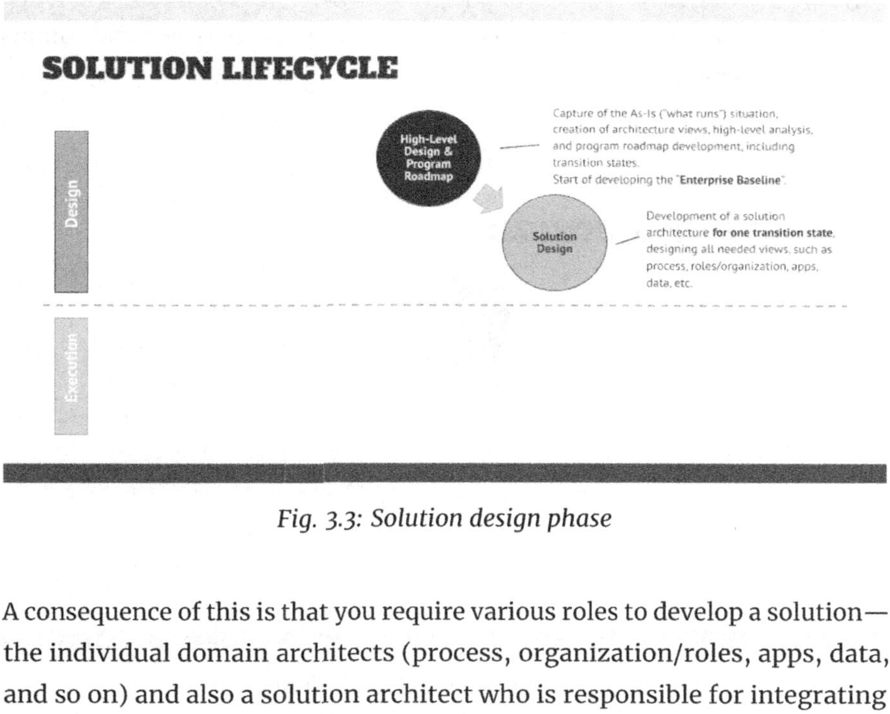

Fig. 3.3: Solution design phase

A consequence of this is that you require various roles to develop a solution—the individual domain architects (process, organization/roles, apps, data, and so on) and also a solution architect who is responsible for integrating the different designs into a consistent solution design. For larger programs, where multiple projects implement aspects of a capability set, you might even need a higher role—the solution portfolio manager.

Once all dependencies—internally and externally—are resolved and the design is approved, it can be moved into the next phase. In addition to this, the solution design phase can also provide input to (the update of) reference architectures, for example, when a new capability and new systems are to be implemented.

Phase 3: Implementation

This phase covers the *actual changes to the organization, systems, underlying technology, and processes*. It is mostly driven by the technical configuration of the runtime systems but also includes the "people side" of the solution—the org change management team that puts the organizational changes in place, coordinates with works councils, etc. It also aims to enable users and admins in the new technology.

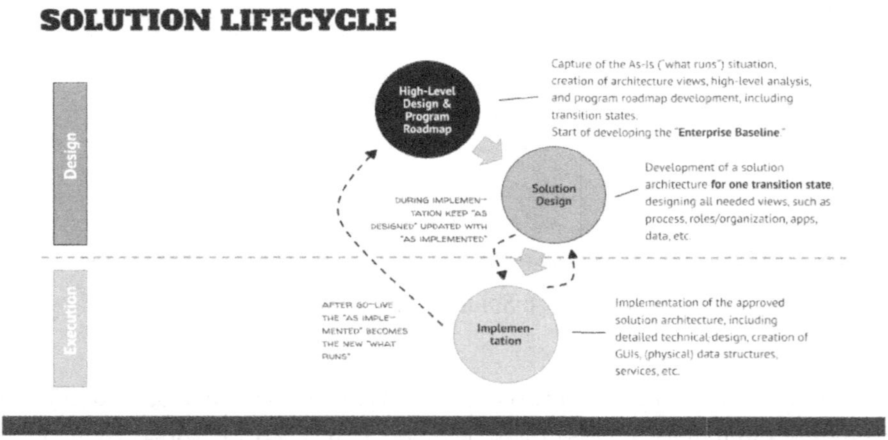

Fig. 3.4: Implementation phase

The architectural contributions in this phase are multiple:

- **Provide guidance** on how things have to be implemented, and update the solution designs accordingly if changes are needed.
- **Support the testing effort** by creating test cases for higher-level tests—remember: every branch in a process is a test case and needs to be

successfully executed.

- **Synchronize/upload solution designs to learning management systems** such as SAP's Enable Now so that the content there reflects the new reality regarding processes, app landscapes, and so on.
- **Update the Enterprise Baseline** with the final implemented design so that future analysis and project definitions can be done from an up-to-date information foundation.

Phase 4: Process execution

The scope of this phase is to implement an *ongoing monitoring of the execution of the solution* on a day-by-day basis. Setting up dashboards for the different stakeholder groups—for example, operators, process owners, management, etc.—will provide real-time information measuring the service-level agreements (SLAs) for the defined processes and solutions.

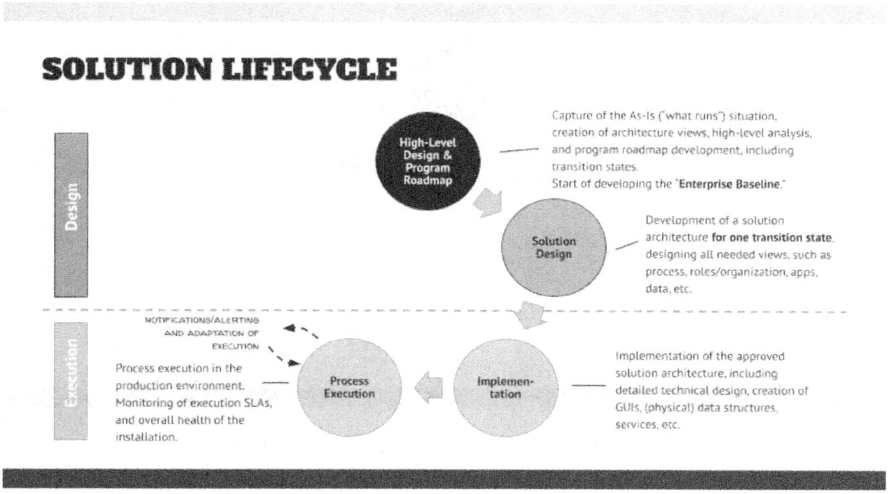

Fig. 3.5: Process execution phase

Aligning KPIs that are visualized by the dashboards with reference content (e.g., ITIL,[8] APQC,[9] or SCOR)[10] will help to compare to benchmarks and get users accustomed to process thinking.

What this phase also includes is the interaction with the execution if an SLA is not met, for example, if a delivery of a product takes too long. Notifications and alerts can be configured to trigger actions as needed.

Phase 5: Measurement and analysis

This phase is similar to the previous one and might also use the same data sources in some cases, but the objective here is to "take a step back" and look at trends and usage patterns that show up after some time. You also might want to take a look at what your users actually do (vs. what might have been mapped out in the solution design phase).[11]

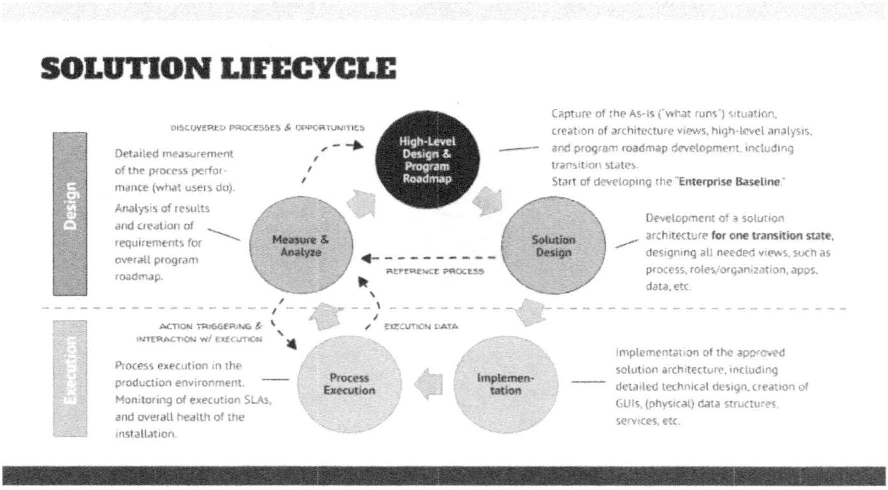

Fig. 3.6: Measurement and analysis phase

The main tools for this phase of the solution/process lifecycle are process mining and task mining (robotics process discovery). Both look at usage and performance KPIs, but process mining typically is fed by log files from transactional systems that produce case IDs, activity names/events, and start/end time stamps. Task mining will have agents being deployed on user machines that record what users do in which program (down to the field level) and then create process models for procedural steps—one level below the process step that you typically see in a process model (a "business-relevant, enclosed activity").

The main purposes of the measure and analyze phase are:

- To analyze the process performance and find improvement opportunities for the next iteration of the lifecycle
- To compare the solution design ("how work should be done") with the reality of how work is done. This is a very intriguing topic not only for process improvement people (business analysts, process owners, etc.) but also for auditors, who will ask both questions very early on in an audit.
- To discover processes in case there is no enterprise baseline for this process. Mining technologies will give you a "good first cut" of the process, and the better tools have a functionality that allows you to transfer the "80% process" to their repository so that you can then define how the work should be done.
- To interact with the execution of the running process by providing functionalities like alerts, notifications, or even running transactions when certain criteria are true—for example, if an overall case or individual activity within the process takes too long. This will allow managing the runtime process better by either escalating and getting a faster decision that still will keep the process within its SLAs or by triggering a re-planning (for example, when a container is stuck in a harbor overseas, a production planner can change the manufacturing plan to produce something else instead of waiting for the container to arrive).

But please note that the interaction with the runtime process is not a "permanent fix" for performance issues of the process. You would rather not manage your process by exception but rather take the insights from the current exceptions and add them to your list of analysis topics or improvement opportunities for the next iteration of the lifecycle.

How do architecture tools support the solution/process lifecycle?

A good professional architecture tool stack provides features that support all solution/process lifecycle phases:

- **Enterprise Baseline**: a central repository that allows the structuring of content as defined in the metamodel; instant impact analysis by using the object orientation in the underlying database; analysis reports for repeatable use; ad hoc queries for quick analysis; simulation; automation of [governance] processes; dashboards for insights into governance ("Are we 5% or 85% done?").
- **Solution Design**: modeling capabilities that allow the creation of blueprints in all needed architecture views; comparison features (to show the difference to the baseline current state); if available, reference content to create consistent designs; analysis features to show the improvements/differences compared to the current state; versioning; if available, implementation tool-specific attributes and workflows for data exchange.
- **Implementation**: synchronization of blueprints with implementation tools (ideally bidirectionally to manage scope); creation of test cases and tracking of those; analysis/reporting for org change management (such as role reports as input for the creation of a training curriculum or communication items); provision of auto-generated views, such as process narratives or step-by-step views; export of content for presentations and other org change management material.
- **Process Execution**: dashboards; "live" connectivity to source systems; actions and notifications.
- **Measure and Analyze**: various analysis types for process and task mining (see above); what/if scenarios; actions and notifications; bidirectional

connection to the design tool.

Figure 3.7 below shows some of these technical capabilities mapped to the solution lifecycle.

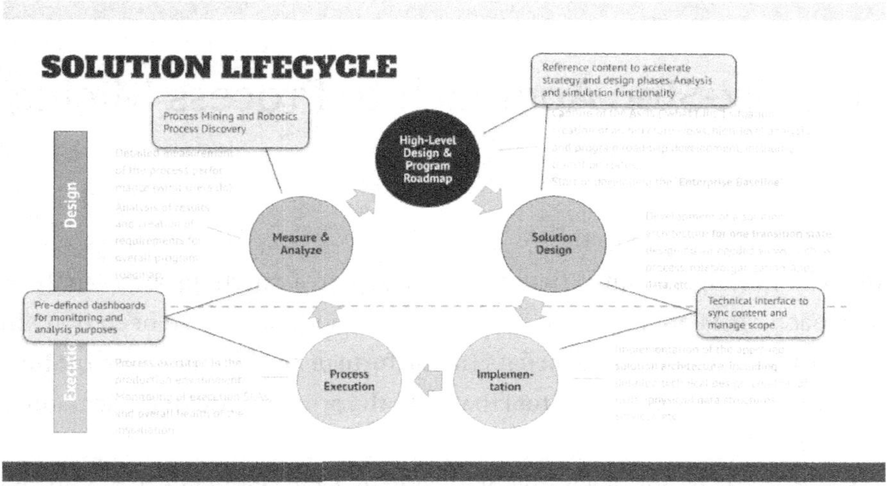

Fig. 3.7: Example tool support in the solution/process lifecycle phases

I think this chapter clarified that architectural work and process mining analysis are not just 'draw and forget' or limited to the design phase of a project. Managed well, it is a competitive differentiation by providing longer-term vision and visualization of a strategy and the necessary information needed for solution design, implementation, and operation.

By setting up the lifecycle in the phases above, a reinforcing effect will kick in that will accelerate the maturity of design and delivery significantly.

4

The Six-Step Approach to Process Mining

Process mining is more than a technology or a single process analysis approach. It is an assembly of analysis methods that have been done in the past but are now based on data (in the form of an event log from runtime systems) and supported by a tool that does the process-oriented calculations for you.

However, it is not as simple as throwing a big data export file on the software, and the tool will 'auto-magically' figure out your process problems. Nor is it a one-step project approach, as some vendors might want to make you believe.

Based on dozens of successful process mining implementations, a six-step approach to process mining projects has crystallized that will be discussed here. This approach will make sure that you get the right process selected and use the necessary data for your analysis so that you can identify the right improvement opportunities that make an impact in your organization.

The six steps in an overview

They say, "Every journey begins with a single step," and that is also true for your first steps in implementing a process mining capability in your organization (and we are looking here at process mining and not task mining or other data-driven analyses, which might follow a similar approach but will change slightly based on the technology used). The following graphic, 4.1, shows the six steps in an overview, and the different roles involved are indicated by the little squares on each step:

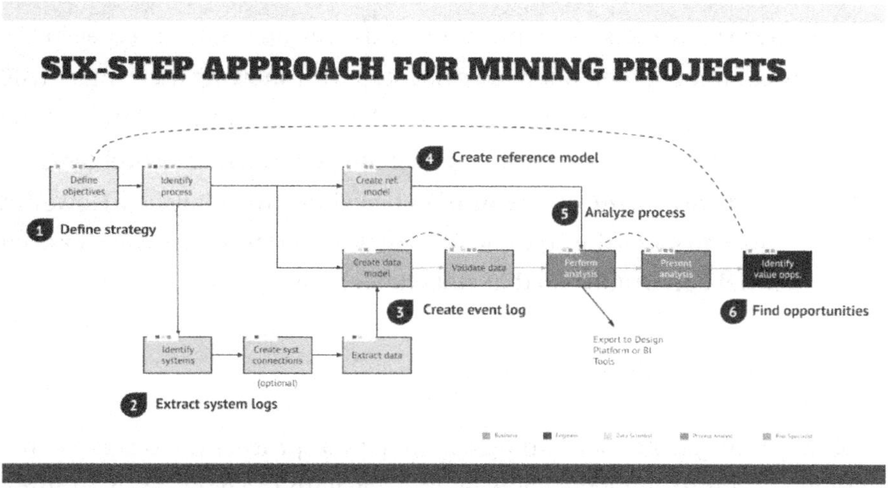

Fig. 4.1: Six-step approach to process mining projects

The six steps are:

1. **Define the process mining strategy**—what do you want to accomplish with this new capability in your organization, how will it be implemented, and what are the first projects and desired outcomes?
2. **Extract system logs**—identify the source systems for each step in the process, create technical interfaces for automated data feeds into the process mining tool (optional), and extract the event logs from the tables in the source systems.
3. **Create an event log**—this step includes the data cleansing and data transformation so that at the end of it an event log that includes information from all source tables can be loaded into the process mining tool.

4. **Create reference process**: one feature of process mining tools is the ability to compare a reference diagram in BPMN[12] format ("How work should be done") with the reality of the event log data ("How work was done"). In this step, the how-work-should-be-done process model will be defined/refined.
5. **Analyze process** is the application of the various analysis types in the process mining tool. But it does not stop with showing the "what" (the discovered process with its statistics) but also the identification of the "why" (and when, where, who, and how) of performance problems.
6. **Find improvement opportunities** based on the analysis. However, this also includes the creation of a (high-level) business case for each potential opportunity so that stakeholders can decide.

Who is involved in this approach?

Process mining is not something that a data or process analyst does in a dark basement to suddenly show up one day with the once-in-a-lifetime breakthrough idea. It is—as always in business—a team sport, and you have to set up a multidisciplinary team to be successful.

- **Business representatives**: the people who are performing the process and/or are responsible for them (aka subject matter experts/SMEs). These are the resources who know the current process with all of its applications, data, "unwritten rules," and performance issues.
- **(Process mining) engineers**—these are the technical people who own the source systems and also the engineers who will build integrations to those systems to enable automated/scheduled feeds into the process mining tool.
- **Data scientists**. Process mining has the new(-ish) quality of being based on the data that comes from the source systems, and you need people who can take raw data and transform it into the tabular format that your process mining tool requires.
- **Process analysts** are the traditional resources who analyze the process

using the different features and techniques in the process mining tool. They also bring in a Lean/Six Sigma background to find the different forms of waste, using statistical knowledge to decide if what you see in the tool is relevant or "just an outlier."

- **Risk specialists**—one big topic in process mining is "compliance" (aka, do my users do what they are supposed to do?), and these are the resources who are experts in it. They also bring the knowledge about risks in the process and the frameworks behind them and can add another perspective to the development of hypotheses.

Step 1: Define Strategy

One of the first things to think about when implementing process mining for the first time is to take a step back and think about what you want to accomplish with this technology and approach. However, there is more because you need to choose the right process for your first steps so that you can build momentum and excitement in your organization.

I recommend going through the following steps and capturing your decisions for future reference and adjustment.

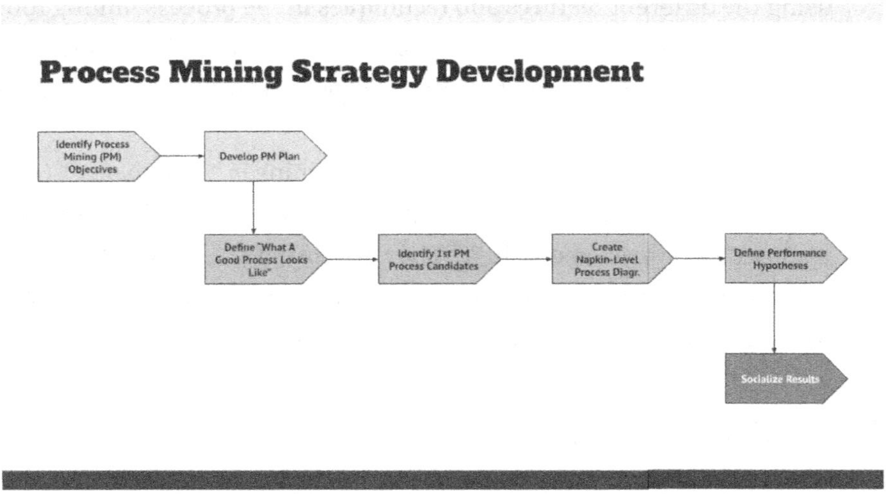

Fig. 4.2: Steps to develop a process mining strategy

- **Determine process mining objectives for your organization.** This should be obvious, but I have seen numerous clients just "trying out" the tools and getting frustrated because they don't give you obvious answers until you have to do the actual analysis work.
  What you want is to define and assemble the list of objectives and then align them to your business's strategy, the running projects, and the organization's appetite for "another new software." Even though you might be excited about the new technology, you may find that other parts of the organization have "transformation exhaustion" and just cannot take in something new. In this case, try to find areas in your organization that are more receptive to this and scale down your ambition—it is better to have something deployed and convert some users into believers than push too hard on uncooperative but "more valuable targets" who will become anything but advocates.
- Which leads us to the next step—you need to **develop a short- and long-term plan** for this capability. Besides the fact that process mining can be an expensive exercise (not only regarding the license costs but mostly

because of the effort and number of people involved), you might want to make sure that you deploy it to your most important processes—the ones where you get a strategic advantage for improving.

Interestingly enough, the majority of examples that I have seen are commodity processes, such as procure-to-pay or order-to-cash. My advice here is to have a nuanced approach to this—if a process is a commodity, you might want to standardize and automate it as much as possible. Process mining might be a monitoring exercise here, and if the numbers you see are "good enough," then you might want to focus on something else. If the process is unique and gives your organization an advantage, then you might want to tweak your analysis to a much deeper level after the first implementation and analysis.

Either way, you need to come up with a plan that balances the benefits for the organization with the cost and the opportunity to build momentum—you need to decide if you want to have "the one" big project or a series of smaller projects that will grow your user base and their acceptance.

- **Determine "what good looks like" for processes** in your organization. This is a crucial step, and we will refer to this a couple of times below.

  There are a limited number of levers that a process can have and that you can influence—for example, the capacity of a process: how many widgets can I produce in a given time (or, for white-collar scenarios, how many loan applications can be underwritten)? A second one is the course of time: how long does the process or a step within the process take? Where are the bottlenecks? There may be other factors like customer experience, cost, compliance, caliber (quality), and so on.

  You need to define a mix of all these levers for your organization. If your focus is on speed-to-market, then you will hire more resources or buy machines and software than if your focus was cost or if the regulators are watching over your shoulder. You might want to document these levers and socialize them in your organization so that you get the buy-in from the decision makers but also the acceptance of the business unit members that will execute the process that you want to analyze.
- **Identify the right process for your first project**: This relates closely

to the previous topics. You want to select a process that has a business impact when you find improvement opportunities (aka not "technology evaluation" but true benefits). You also don't want to tackle the "big hairy beast" as your first project because you will learn a lot about the process of doing process mining projects—these six steps here are a good start, and you also want to have a quick turnaround of results in a few weeks. Your stakeholders will lose excitement if you take months to present them results that they then might find underwhelming—not a good start to build momentum.

- **Create a "napkin-level" process diagram** of the process that you have identified. It would be great if you already had a mature BPMN diagram of the process in scope, but at this point in the game you don't need that. What you want to have is a level of fidelity that allows you to identify the steps of the process, its logic, and maybe some exceptions that will help you in identifying the systems of record for each step.
  This will also help you determine who you need to involve from a business SME perspective and check the feasibility of a potential timeline. If you cannot get an SME for your process, then you should choose a different process.
  In step 4 you will take this rough process and create a more detailed version of it in BPMN format that depicts how work *should* be done in more detail.

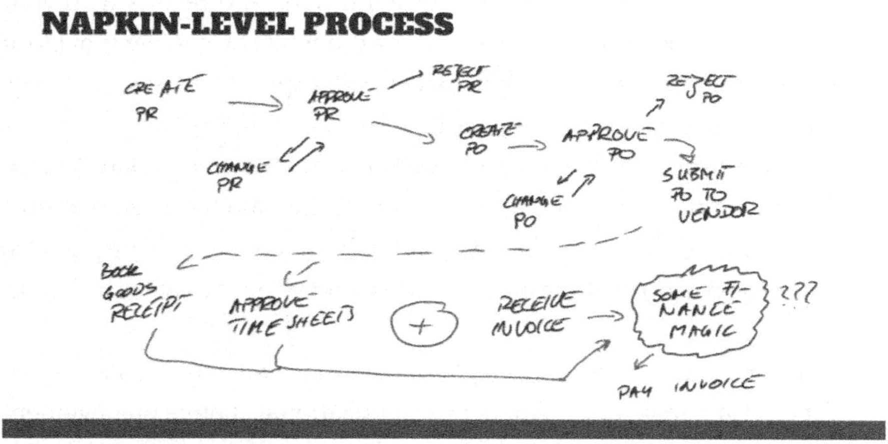

Fig. 4.3: Initial "napkin-level" process diagram

- **Define hypotheses for performance issues**: One misconception of process mining (which is sometimes communicated by the tool vendors) is that you just have to connect the data to the tool, and you're done.
  It is true that you can get some good analysis results from the mandatory information in the data set and the upload of a reference process, but the more in-depth insights come from performing a detailed analysis that is aligned to hypotheses. What is always wrong in your process execution? What are the reasons? What could be resolutions for solving these problems?
  What you want is to have a list of hypotheses and assumptions that lead your data collection (step 2) and process analysis (step 5). You want to have the needle in the haystack defined instead of just pointing to one (or more) haystacks without instructions for what to find in them.
  Remember: you can always extend your analysis, and if your initial hypothesis-based analysis does not bring the "breakthrough results," then you can always extend the hypotheses and continue your analysis in a future iteration.

- **Socialize the results with your stakeholders**: As mentioned above already, you have to socialize your definitions and results with your stakeholders. Keep in mind that you have different stakeholder groups for different topics—your decision-makers will approve budgets, but everyone will be interested in things such as "what good looks like" definitions. On the other hand, you might have much smaller stakeholder groups when it comes to the individual processes that you have selected on a working level, but you definitely want to have their leadership included in the main aspects of the processes ("informed" or "accountable" RACI relationship).[13]
  What you want to create is a stakeholder map, which is a topic in itself for a future book, and engage the organizational change management professionals who have this method in their standard toolbox. They most likely also have experience with the individuals in your organization and know what their preferred communication style is.
  You also want to plan communication activities and align them to your stakeholder groups in the program to say, "This is relevant for all stakeholders, but this other content is only relevant for this subgroup."
  The main takeaway is that you always want to have your stakeholders engaged and show progress that is meaningful *for them*.

Step 2: Extract System Logs

Now that you have identified the process in scope and articulated what you want to analyze (in the form of the hypotheses), it is time to go deep into the systems to get the right data that will support the process mining analysis.

Keep in mind, if you cannot find the right data, it might be the right time to raise a flag and go back to step 1 (and the stakeholders) and assess if you have chosen the right process. It is better to go back to square one than to occupy valuable time from all people involved and have a mediocre result that disappoints your stakeholders. Remember: you want to build momentum and excitement for this new analysis technique and technology.

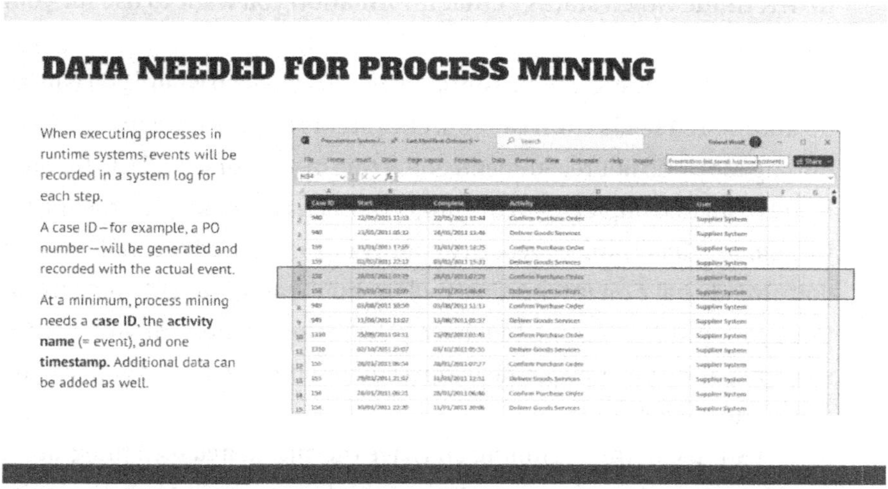

Fig. 4.4: Data needed from runtime systems

- **Identify systems of record and additional systems** needed to give context when analyzing the hypotheses: Most likely you will have more than one source system for your process (for example, a procurement system, a warehouse system, and a financial system in a procure-to-pay process) that might or might not share the same identifiers for the individual process instances/cases.
  In this step, you need to do three things: *identify the source tables* in the systems, *determine which columns you need* to include (based on your hypotheses, and most likely a DB administrator will have to write a custom query for you), and *determine the output format* of that data—do you get one big file, or do you get multiple tables based on the type of information that you will need? The general rule of thumb would be that you want multiple tables—the actual activities and their timestamps, and then additional attributes for each step in this activity log in one or many files. In step 3 you will combine them in a way that makes sense.
  One thing you also need to understand is if data will be changed in different systems during the execution of the process, and you need

to determine which status of that information you want to use for your analysis.

- **Build technical integrations (optional)**—this is a step where you will set up automated feeds from the source systems to your data modeling or mining tool. I do not recommend doing this in your first projects because it adds a lot of complexity to the project, which can make the timeline for it explode ... just think of security approvals or the used technology—do you already have an integration tool in-house and available resources? Does your process mining tool have a dedicated adapter for your source systems, or do you have to develop a custom interface (and who can do this), and so on?
  Keep it simple in the beginning and save the automated feed integration for a future version of your analysis. Initially, you want to show results quickly.
- **Extracting system logs** is the last step. Here you will have to decide how you organize your results, where to store them, and which naming convention you apply to the files.
  This is no problem if you have built a technical integration but can be complicated if you work with exports from databases and other systems. For example, I worked on a project with a bank that gave us the online transactions of their customers' investment changes, which were recorded from three different systems. Besides that, it was challenging to connect the three sets of records. Once we had created the initial data model based on one day, they gave us the records of one month (90 files instead of the original three), and we were happy that the data modeling tool that we used was able to simply switch from "read this file" to "read all files in this folder," and it automatically removed redundancies, etc. Long story short, you should have a plan on how to organize your source system logs and plan it in a way that makes it easy to expand if you get a mass export of data. What you should avoid doing is artificially reducing the amount of data that you will get—the more you have (and I am talking about hundreds of thousands or millions of records), the better your analysis will be.

Step 3: Create Event Log

This step is the most crucial step in the whole approach. Get this wrong, and all steps downstream will be wrong, and your recommendations might be invalid.

Use a graphical tool for creating your data model

When approaching the creation of even logs, there are various tools out there, and typically the process mining tools ship with some form of transformation capability. Those are most likely in the form of nerdy SQL queries or other forms of code and not necessarily stakeholder-friendly. Even if you have the necessary skills for this in-house, the use of "coding" might be a turn-off for your business stakeholders and also might be challenging to learn for them, and one of the objectives that you might have defined in Step 1 might be that you want to roll out process mining in your organizations.

I recommend using a graphical data modeling tool for this step, and there are commercial tools like Alteryx[14] as well as open-source tools like KNIME[15] available. These tools also come with training and certification offerings, as well as good documentation/videos and (maybe most importantly) an active user community that can help you to figure out how to set up the transformations. This will simplify the rollout of the tools significantly.

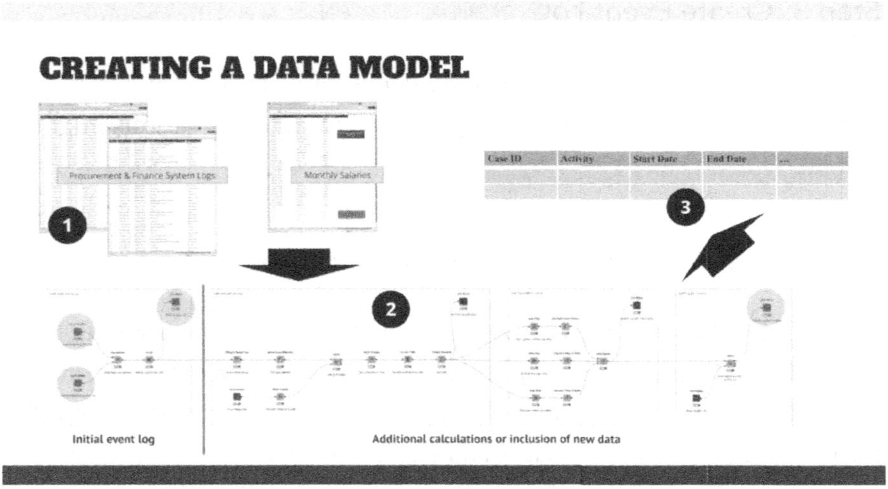

Fig. 4.5: Create a data model

Create the data model

Creating the necessary event log, in the format that process mining tools require, can be a challenging exercise, and you need to be aware at every step of your data transformations that you do not accidentally lose or add data to the final table. Your end table must have at least the first four columns shown in the graphic above: *Case ID* (so that you can trace the events by their timestamp[s]), *Activity Name*, and *Timestamp(s)*—ideally start and end timestamps, so that the tool cannot only calculate how long it took to get from step A to step B but also how long it took to perform step A. The tools will then give you additional statistical calculations on them like average, mean, max, and min values.

To create a data model in a graphical tool like KNIME you have to:

- **Import event logs into data tool**: KNIME has tons of connectors to flat files, databases, etc., so that you can read almost any file type that you throw at it. Once you have it in the tool, you can see the columns and

determine if the data was imported correctly.
Obviously, you can import more than one source file or connection, and you can also create your own (reference) tables that you might need for calculations, as well as export the final table(s) in all common formats.

- **Cleanse your data**: This is the first crucial step, and it includes a mix of table changes, like removing columns that are not needed or changing the format of a column, to row edits, like filtering out events that are not applicable because they include cases that already were in progress when you took a snapshot. Or you have to do cell-based edits like harmonizing strings or date calculations, for example, if one source system stores timestamps in 24-hour format, while another one uses AM/PM format.
- **Create an initial event log**: focus on the base information that you need for your process mining tool and upload a first version of your event log into it. What you want to see is if it a) loads, or you need to tweak the table (date formats are always troublemakers), and b) if the result that you see makes sense at first glance.
  Go back and check your data model if your initial event log is not correct.
- **Fine-tune your data model**: this is where the fun starts. Go to your list of hypotheses and see what questions you can derive from them to be able to reject or validate them. Check your data to see if you have all the information available to answer the questions and determine if you have to calculate additional metrics. A little further below, we are discussing the example of an incident management process—you might have to calculate which percentage of cases are escalated to a Level 2 support organization.
  During your process analysis, you will see plenty of iterations of changing your data model once you have presented your initial results to the stakeholders and collaborators. Therefore, you might become very familiar with your data modeling tool and/or the calculation features that the process mining tool might have.

Validate your data model

Once you have created your initial data model, load it into your process mining tool and schedule a data validation session with your stakeholders. The main objective of this session is to validate the data model itself and not so much the results that the tool shows.

I typically start this workshop by reminding the participants that this is a working session and the main aspect is to ensure that the data model is correct. Then you should walk them through your data model, and here it is very helpful that you use a graphical tool and not just a series of SQL queries or Python scripts—those are typically not the skills that your stakeholders have, which results in them "checking out" because it is just 'blah blah' for them.

An added benefit of using a graphical tool is that they are "self-documenting" and allow you to add descriptions to the workflow nodes and also group nodes into logical chunks of your data transformations—up to the point that you create collapsible and reusable groups that can be referenced in other workflows. Very helpful when you create data models for similar processes or when cleansing your data.

One reaction that I have seen over and over in projects is that stakeholders become defensive and claim that your data model is not correct and therefore the spaghetti diagram that they see in the tool cannot be correct. This is the time to ask them to give you some case numbers and check the information (e.g., timestamps or the sequence of steps) in the source systems like an ERP or CRM system.

You might find some errors, which is good and most likely a result of assumptions that you and the stakeholders have agreed on. Analyze where the error in your data model might be and iterate on the data model.

It is crucial that you get the OK from your stakeholders that they believe that your data model is correct; otherwise, all other steps are moot. Have all participants state that they agree with the correctness of the data model before you release them from the workshop.

Step 4: Create Reference Model

As mentioned above, there is an analysis technique called conformance check that compares the runtime data in the event log with the reference process that describes how work should be done.

This is the time for two things: first, to create a proper process diagram in BPMN notation, especially when you just have a "napkin-level" process created in step 1. But second—and maybe more important—you should review the process diagram to update information that might be outdated since the last update/review or remove inconsistencies in the process flow. The outcome of this step should be that everyone agrees that this is how the work should be done under the current constraints, like existing IT support. If you already notice ideas on how to improve the current state, capture them either as a hypothesis for the next step 5 ("What happens if we remove this activity completely? This would take out one system from the process, and we could retire that app" could be a valid result), or directly as an opportunity for step 6, in which you would create the business case for that idea.

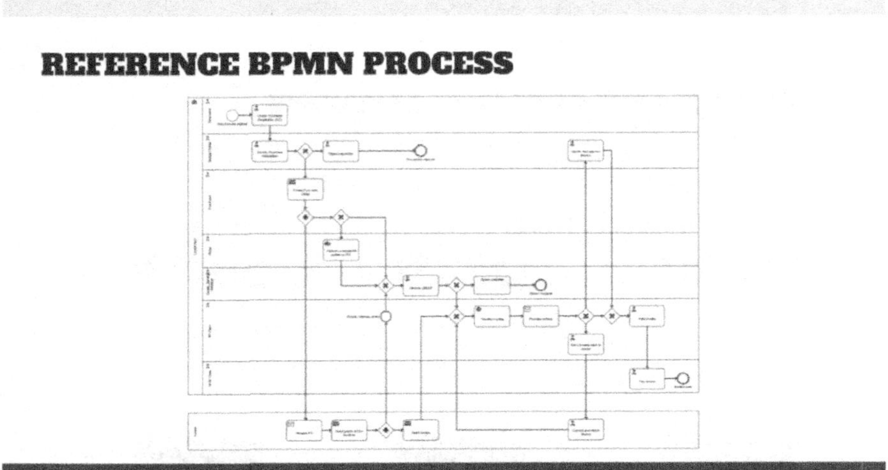

Fig. 4.6: Reference process in BPMN notation

Please note that most likely you will just have the "happy path" being documented and maybe a few exceptions—this process is created by what the SMEs know or want to tell you and is therefore highly subjective—which is OK at this point. You will be surprised to see the creativity of users to bypass "inconveniences" and "process gaps" when running that check.

Step 5: Analyze Process

This is the main step that everyone has been waiting for—you finally take a more in-depth look at the process that was "discovered" by the software based on the event log.

When doing the analysis, which consists of many analysis techniques, I suggest doing this in two steps—a high-level analysis and a more detailed analysis, which will then go into the different hypotheses from step 1. In either of those steps, though, keep in mind that the tool will just show you the "what." There might be no context in the data or any additional hint. You, as the analyst, will have to look for the why, how, who, and when, and your hypotheses will guide this search.

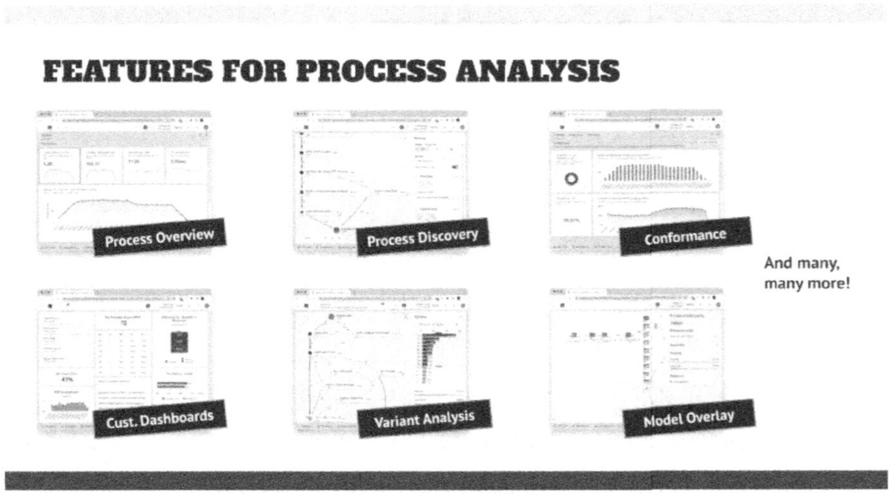

Fig. 4.7: Tool features supporting process mining analysis

High-level process analysis

In the high-level analysis, you typically look for these items:

- **Overview dashboard**: This might not be relevant for your first process mining project, but it is a good idea to define standard KPIs for your processes that stay the same for every process that you will analyze. Some tools already have a pre-built dashboard for this, but you might want to customize this and add metrics like number of cases, average duration, degree of rework, degree of parallel work, distribution over time, etc.
  Having this standardized overview dashboard for every process brings focus to your process analysts and process owners on what is important for you as an organization and allows an "oranges to oranges" comparison.
- **Process discovery:** Process mining tools "discover" processes based on the sequence of the steps in the event log and their time stamp. This can show many interesting observations, for example, with which events the process starts or ends, or which loops might show up. In these cases, you will have to see if it is really a process path that you are looking at, or if your data set is "skewed" because you have not included all events of already completed cases (because you defined a cut-off date), or if some cases simply have not completed yet.
  You might want to change your data model for the first case and highlight the "in progress" cases in your analysis.
  Please note that some process mining tools allow you to export a process to their repository. If your tool can do that, I recommend applying the 80/20 rule, selecting all variants that cover 80% (-ish) of all cases, and exporting them as a BPMN diagram. In your repository, you can then do a model comparison with the diagram that you have received from your client or created in step 4. You can even go back to step 4 and use that input to refine the reference model that will be used in the conformance analysis below.

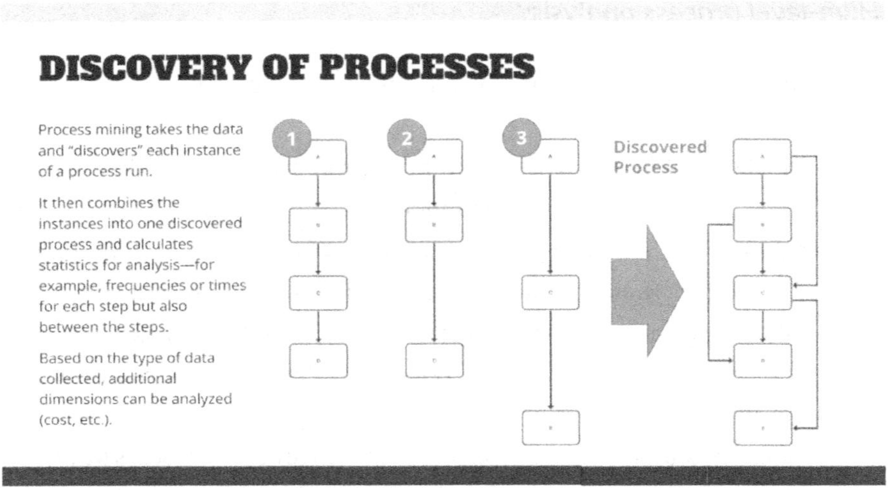

Fig. 4.8: Process discovery

- **Frequency Analysis**: This is the first major question you are asking—how many cases go through a certain path? This will determine if you will focus on other aspects in later analysis (is it critical that this group of processes take significantly longer than others if the total cases are just 0.2% of the overall number of cases? Most likely not), and you can identify the "happy path," aka the most used path, and compare it quickly with your reference process to find exceptions and potential areas of further investigation.
- **Time Analysis**: Besides the question "how many?" the second most important question is "how long?" (Remember the levers of a process from step 1.) In this analysis, you look at the time that certain processes take—this can be the overall process execution time or the time a single task in your process takes.
  Compare that with the SLAs that you have captured in step 1. Do you see a discrepancy? Is that a general problem, or can you isolate certain conditions for the discrepancies—for example, certain products or vendors in a procure-to-pay process?

- **Bottleneck analysis**: This is a variation of the time analysis where you explicitly look at the delays that are showing up. Often, this is organization-related (you just don't have enough people, and the cases pile up), or a lot of rework needs to be done, which slows down the organization that is working on the cases.
  Note these numbers and use them in the process simulation detailed analysis below to figure out if and how this problem can be solved. And every so often the problem is not "in the chair," but your means of transportation is outdated, i.e., you have things pile up in email inboxes instead of using a workflow system that has dynamic routing of cases.

Detailed process analysis

The detailed analysis now can drill down into more aspects of the why, how, who, and when by using these techniques:

- **Filters**: Most likely, your initial discovered process does not have a clear picture or observations presented to you. Go to your hypotheses and use the filter functionality of your tool to dig deeper into aspects of the process—for example, filter by assignment group of the Level 2 support in an incident management process to see if some L2 groups perform better or worse than others.
  Save these filters globally for your project and come up with a naming scheme for the filters so that you can easily replicate the view in a dashboard without having to click around for five minutes every time you want to revisit your analysis.
- **Variant analysis**: Variants are "walkthroughs" of cases through the process, as shown in the graphic below. What you want to look at are the most used variants, and you most likely will see a "long tail" distribution where the majority of all cases are in the first 5–10 variants and then the distribution tapers off until at the end of the variants you will see the "one-off" cases.
  Most likely, you would rather not focus on the outliers except if there is a

specific reason for it—for example, you have cases that have a very high cost, and you want to analyze why that is—but rather apply the 80/20 rule again and look for the majority of cases.

In some tools, you can also overlay the BPMN diagram/reference process with the variant data (filter by variant), and you can more easily identify where the variations occur—sometimes you might even see that there are multiple exceptions in a variant.

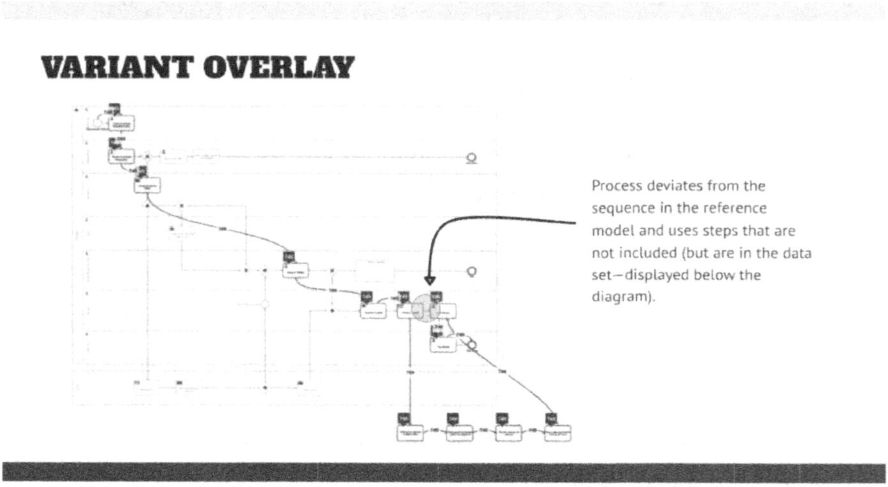

Fig. 4.9: Variant overlay in a process mining tool

- **Conformance check**: This is the analysis where you load your reference diagram into your process mining tool, and it compares the data set to your diagram. It will show you the differences between the two and the associated statistics for each, like frequencies or times.

  Typically, you will get multiple results from this analysis—for example, an overview dashboard of your overall compliance to the reference process over time and absolutely (as a percentage), which you can filter

on. Or you get a detailed list of conformance violations, such as wrong start or end event or sequence of steps, which you can immediately use as the starting point for more analysis and/or identification of improvement opportunities in step 6 of the six-step process later.
In my opinion, this is the "killer analysis" to get risk people on board with process mining because the third line of defense (independent validation via audits) is extremely interested in seeing what process participants *really* do, and process mining can give them crucial insights that will help them in their auditing efforts.

- **(Custom) dimension/metric analysis**: Custom dimensions and metrics are the detailed pieces of information in your data set that you will assemble to validate/reject the individual hypotheses.
  My recommendation is to build a single dashboard for each hypothesis and avoid creating a "one-size-fits-all" dashboard if the hypotheses are not related to each other (even then I typically recommend individual dashboards because in some tools you have very granular permission controls so that you can specify a subset of analyses for certain stake-holder groups). This avoids them being overwhelmed by what you have found.
- **Role analysis**: This analysis changes the perspective from the steps in the process to the participants. Some tools already provide this overlay out-of-the-box, while for others you might have to load your data model so that the "activities" in the tool are the roles that you have in your data set. This will show the main interaction between roles/teams that might give you additional insight into the context of the process—for example, that certain roles/groups are systematically excluded from the process, which then creates delays.
- **Data Flow Analysis—by system**: For this analysis you want to look at two perspectives—one is the sequence by application (to find potential missing automated interfaces where a human has to enter data as the "swivel chair interface," which always has the danger that information gets lost or tweaked, which might create additional bottlenecks because further downstream in the process there needs to be an additional verifi-

cation/correction step).

The second is the analysis of the degree of automation of the process. For obvious reasons, the higher degree of automation that exists, the better the data quality will be that is used in the process. This, and management of the workflow, might have the largest impact on the performance of your process.

- **Case analysis**: While until now, you have looked at patterns of processes, the process mining tools also allow you to look at individual cases and how they flow. This will help you find the specifics and context that will help you to interpret what you see in the higher-level flows (e.g., in bottlenecks or variants).
- **Prediction**: Some process mining tools have predictive capabilities using AI/machine learning to project how a case in progress will end with a certain probability. This information can also be used for triggering actions that some tools allow—for example, if a container is stuck in the harbor in China, the prediction can give an estimate on what it learned from the data set about how long this typically takes so that one can change the production planning to avoid delays.
- **Simulation**: Typically, process simulation is a separate component of a business process management suite, and only rudimentary what-if functionality is included in a process mining tool. I highly recommend taking a closer look at a full-fledged simulation tool in the context of process mining projects because the functionality will help you to detail your improvement proposals.

  In general, you want to take the metrics from process mining and use them to create a baseline simulation, which will give you a starting point for comparisons of simulation scenarios by calculation of resource utilization, waiting times (static, dynamic), etc. You can then fine-tune your identified opportunities later to see if, for example, the change in the number of resources will remove bottlenecks, and then furthermore if you can remove people from downstream activities and what impact that might have ("Does this remove the bottleneck upstream but create a new bottleneck downstream?"). Or you can see the usage of other, non-human resources and optimize your recommendations toward the better utilization of those, say, in a production process. Or you can simulate what the reduction of exception paths in a process would have on the process or have the simulation tool calculate the optimal use of resources and apply a schedule to it.

The possibilities are—literally—endless, and you will have to decide when "good enough is good enough" for your simulation analysis if you would rather not have the overall project schedule explode—the criteria for "what a good process looks like" from step 1 will give you guidance on this.

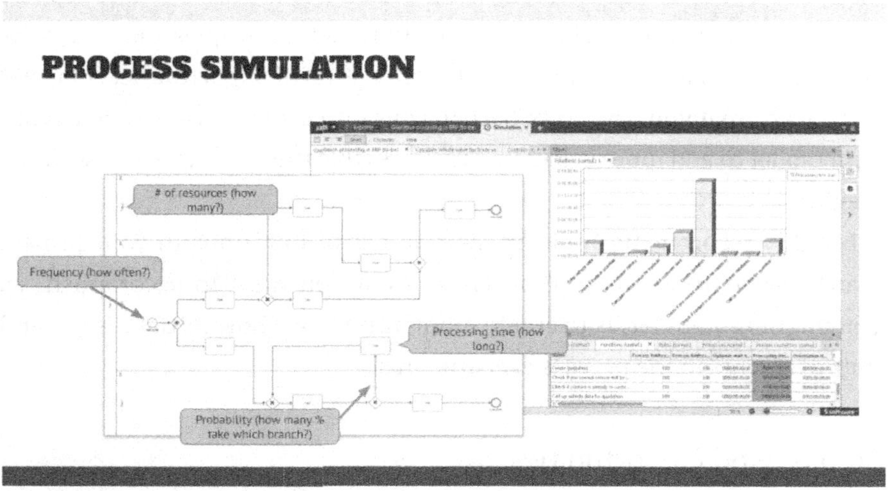

Fig. 4.10: Process simulation example

The possibilities seem to be overwhelming for beginners, but the more process mining projects you have done, the more you will realize what the overall behavioral patterns in your organization are. Just remember: each of those analyses should be aligned to your hypotheses, and you should create one dashboard for each hypothesis in the tool.

Present analysis

When you have done your initial analysis, invite your team for a presentation of the initial findings. What you typically will see is that there will be plenty of questions coming up or attempts to explain why certain patterns show or why certain org units show a (lower) performance. Note these remarks and add them to your hypotheses for the second round of analysis.

What you will also find is that a lot more questions might come up. For example, in one project in which we analyzed the IT incident process of a bank, we looked at the number of cases that had been escalated and showed that certain software/hardware had more escalations than others. This is a finding in itself that has a value, but then the question came up on how many of those escalated incidents were escalated to the third level of support (typically the vendor) and what impact that had ... The questions from the team are invaluable for your future identification of opportunities, but that also means that you have to go back to your initial analysis and then to your data model to add more calculated metrics (like the escalation of escalated incidents).

I highly recommend limiting the number of iterations in your project plan/charter to not more than two. What you want to accomplish, as you might remember, is to create momentum and show initial results and improvements in the process execution fast.

Step 6: Find Opportunities

After the analysis, the identification of improvement opportunities follows—even though by now you should have a good understanding of where they might lie: your data tells a story, and the collaboration with your business representatives gives you the necessary context for your recommendations.

The development of recommendations should happen in two steps that go together and allow your decision makers to choose the recommendation that you have prepared for them.

Initial list of recommendations

Capture all areas of improvement in one place and give a high-level benefits calculation for each. This can be high-level and be like, "If we reduce the number of cases that go down that exception path by 10% [by updating our outdated business rule], we will save x hours of work that will save $y." You don't have to get this right to the last cent, but you should have a good estimate of the benefits of that recommendation—either in dollars or a business outcome, like improving your product production capacity and therefore more sales.

Then add additional criteria to this list of implementation-relevant information. Your greatest idea is worth nothing if you cannot realize it. When you do this exercise, it will become clear which are the "high-dollar tickets" on your list.

Then go to your work group and present the complete catalog of potential improvement ideas and determine a list of 3–5 recommendations that you want to further detail out for your decision makers.

Create business cases

For each item on your list of to-be-presented improvement recommendations, you have to do two things so that the decision maker has a clear picture of the context of your recommendation:

- **Describe the impact of your recommendation on the levers of the process**. Remember the exercise to define what a "good process" looks like? Well, there is a limited number of levers that a process can have, and in the graphic below you see some of them (feel free to adapt them and add more if needed, like "Caliber" for "Quality" if you want to be cute and continue the list of terms starting with a C—but try to stay below 10 items to not make it too complex for your stakeholders to follow).
  Each improvement opportunity will change the sliders on the graphic, and you need to understand how one change impacts another one. Say, for

example, your recommendation is to hire 10 new people for your process, and you have figured out via simulation that you cannot just reassign existing resources and train them—this will increase your cost (negative) but improve your capacity and course of time (positive). However, since you add more individuals, your compliance might go down (negative). Depending on your ideal process profile, these might be good changes that you are willing to make.

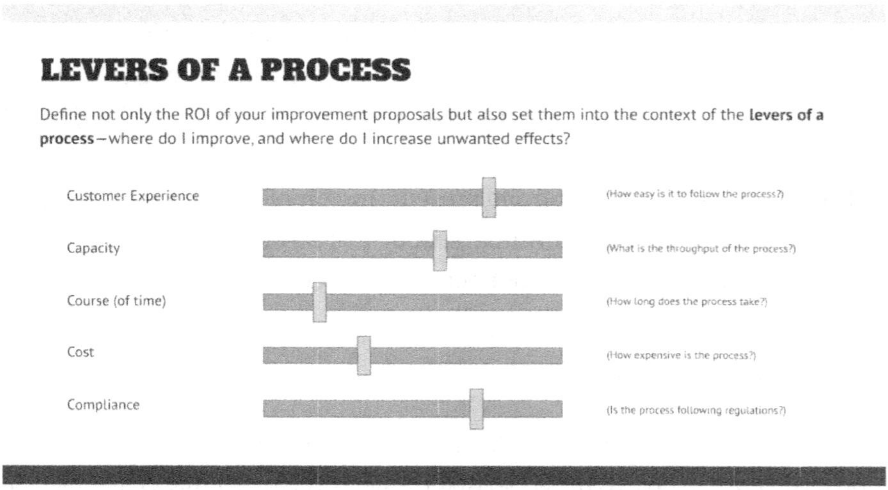

Fig. 4.11: Levers of a process

- The second thing you need to do is to **calculate the return on investment (ROI) for your proposed changes**. In my example above, you would reduce the time and increase production by 50 pieces of your widget per day. This makes about 12,500 additional widgets that you sell for $500 on top of your current sales number—an increase of $6,250,000, while you pay your 10 resources $80k annually ... your potential benefit now is $5,450,000 (6,250M-800k).

You can do similar things by calculating saved hours by avoiding rework or exceptions or any other metric that you find and that makes sense for your process.

- The third item on your business case creation is to **list the assumptions, implementation/changes, constraints, and their cost to the opportunity**. This feasibility check is crucial to get stakeholder buy-in since they want to know how you intend to bring your idea to life—is it a single project or a multiphase project that has significant dependencies on other projects that your organization is currently going through?
  If your idea takes too much to bring a benefit for your organization in a feasible time, you might want to list your idea a bit further down and bring other "low-hanging fruit" items to the top of your list.

Presenting to your stakeholders

The final step is to get the buy-in from your decision-makers. My recommendation is to set up a formal presentation in which you walk them through what you did and what the scope of your project was. Then describe the highlights of what you found in the analysis (an abstraction of the analysis review presentation that you did earlier in step 5) and make clear how these findings relate to the decision-makers, which means you most likely will have to simplify your results (and not nerd out) and speak in business terms and not in the terms of the tool.

Then you come to the highlight of your presentation—the improvement opportunities that you have found. Mention that you found more opportunities, but you applied defined criteria for prioritization of them, so you will just present the 3-5 most important opportunities.

Present each of them by listing the three areas above (levers of process, ROI, and implementation feasibility), and make it simple for them to follow. If you can, try to give the decision-makers options so that they get the feeling that they not only have to make a black-and-white decision but can also add their "stamp" to the future project.

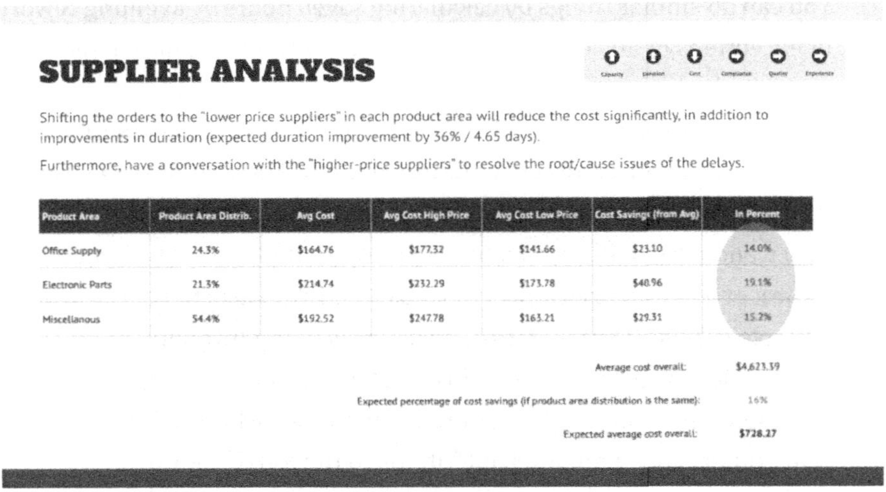

SUPPLIER ANALYSIS

Shifting the orders to the "lower price suppliers" in each product area will reduce the cost significantly, in addition to improvements in duration (expected duration improvement by 36% / 4.65 days).

Furthermore, have a conversation with the "higher-price suppliers" to resolve the root/cause issues of the delays.

Product Area	Product Area Distrib.	Avg Cost	Avg Cost High Price	Avg Cost Low Price	Cost Savings (from Avg)	In Percent
Office Supply	24.3%	$164.76	$177.32	$141.66	$23.10	14.0%
Electronic Parts	21.3%	$214.74	$232.29	$173.78	$40.96	19.1%
Miscellanous	54.4%	$192.52	$247.78	$163.21	$29.31	15.2%
					Average cost overall:	$4,623.39
					Expected percentage of cost savings (if product area distribution is the same):	16%
					Expected average cost overall:	$728.27

Fig. 4.12: Example recommendation—including impact and improvement calculation

The ideal outcome is that you get a verbal approval for your proposals and can start formalizing the projects per your organization's process.

Summary

Process mining can be complex, and there are numerous moving parts that you have to align to get to the point that you have actionable improvement opportunities. But it is also following a relatively simple process that we looked at in the six steps above—if you follow the points in each of the steps and think in iterations instead of a single project, then you will create the momentum that you want for your program.

In general, you might want to keep the 10 points of the following graphic 4.13 in mind when you get started on your process mining journey.

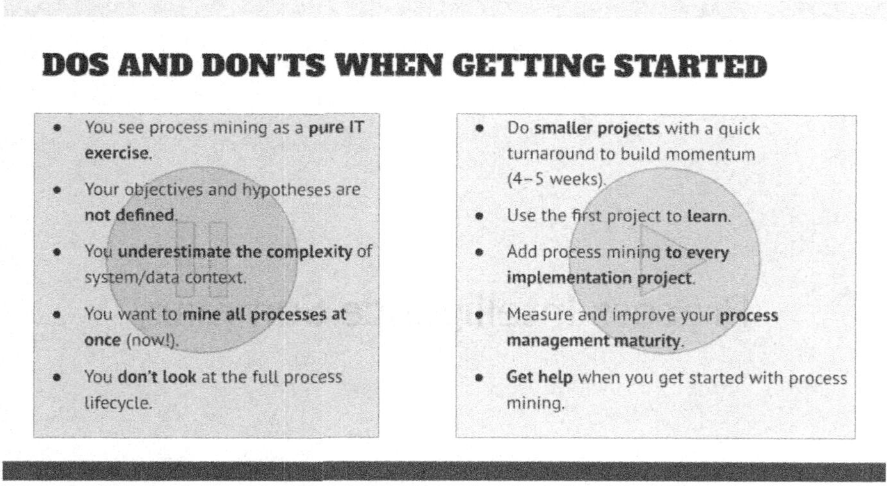

Fig. 4.13: Dos and don'ts when getting started with process mining

If you have additional questions or suggestions, don't be shy to reach out and start a conversation—we are all in the same boat here :-)

5

Process Intelligence Summary

This first section of the book has set the stage for the rest of it, and we looked at process intelligence and how the "new" mining technologies fit into the confluence of data and human-driven analysis. We also looked at how process mining fits into the bigger solution/process lifecycle. Finally, we took a look at the approach to planning and executing a process mining project while setting the foundation for a larger rollout of this method in your organization.

1. Process mining (and task mining) belong to a set of analysis methods that are right between a pure data-driven approach like BI and a solely human-driven means of process analysis. Neither of the two extremes is sufficient, and when you exclude humans from the data-driven analysis, you will miss the context, and the quality of your results will suffer from it.
   In addition to this, the analysis is just one part of the three things that you need to get in order: visibility that allows you to see where things need to be improved, using data-driven analysis with human context to identify improvement opportunities, and then implementation of these improvement ideas in solutions. These solutions can be system implementations, organizational changes, process changes, etc. ... nothing out of the ordinary here.
2. Process mining is one of the main enablers within the larger lifecycle of

process management. It provides means to measure process execution, determine the degree of compliance of how work should be done and how work is done in reality (and make auditors happy), and can help discover processes that are currently not formally mapped.
However, it is not enough to just stand up mining and have automation. You need to have a bigger approach to managing processes, and that includes a repository as well as solution designs before it comes to (system) implementations and running new processes in a runtime tool.

3. When it comes to bringing process mining into your organization, you will have two aspects to consider: how do you position this exciting technology in your arsenal of tools and implement it, and how do you execute your actual process mining projects?
4. The approach that I recommend for the latter is the six-step approach to successful process mining projects that we will follow through in section III of this book. It includes the (process mining and process) strategy, the technical aspects to get runtime data from the source systems, creating a data model as the centerpiece of any process mining project, the development of a future-state reference model, the analysis itself, and the identification and recommendation of improvement opportunities.
   If you follow these six steps, you will get good results. And always remember that you should do several small projects instead of one large project so that you can create momentum—"Momentum is a deliverable."
5. While going through the six steps, keep in mind that you not only work on the content of the process in scope. Process mining is to a large degree managing the stakeholders involved. They give you the content context but can also make or break your project success if you do not explain well enough what you are doing, why you are doing it, and how to interpret the results of your activities.
   Nothing is worse than presenting the discovered "spaghetti diagram" of your discovered process for the first time without explanations and the

main stakeholder tells you that this must be wrong because you screwed up the data model. Always set the context of your interaction right, and rather over-communicate than disappear into a dark office for days and weeks and not talk to anyone.

6. Especially in technology-oriented organizations, you will have to work on some process management fundamentals with your clients—what does a good process look like? How do you prioritize one solution over another? Why do you need to model a process, and what is that fancy and complex notation when the process in reality is so straightforward? And so on, and so on.
   Process mining needs to be embedded in the larger discipline of business process management to realize its full potential.

II

Getting Started

6

Getting Started Introduction

Dipping your toes into a new ocean of technology and techniques can be intimidating, and process mining is no different. But it is also "just new," and many people have mastered this problem and created immense value for their organizations. So don't be afraid.

What's in this section?

This section covers the "things around" the topic of process mining that you need to consider when making your first steps into that new ocean. It covers:

- A chapter that will bring more clarity to the gobbledygook that you might hear around mining technologies
- Guidance on criteria for what to look for when selecting a process mining tool
- A description of the buying process—it is enterprise software in the end, and the enterprise software buying process can be "interesting" if you haven't been included in one yet
- Some short(-ish) thoughts on how to implement the tools and the approaches in your organization

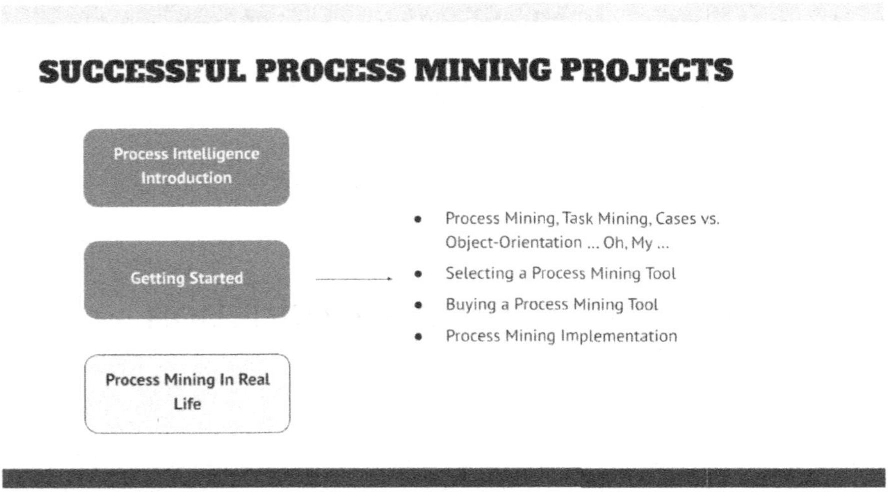

Fig. 6.1: Getting started structure

7

Process Mining, Task Mining, Cases vs. Object-Centric Mining ... Oh, My ...

Looking at process mining is similar to looking at the architecture discipline in general—aka business process management, enterprise architecture, business architecture, or however people like to create their niche. There are tons of abbreviations, concepts, and other things that a newcomer doesn't know what to make of. And why do folks in process mining give themselves fancy titles like "data scientists"? Do you need to wear a lab coat to do things?

In previous chapters we already took a look at the main aspects of process mining, but I'd like to take a moment here to elaborate on some aspects that might not only confuse people but also might have an impact on the following chapters.

Process Mining versus Task Mining

Many people think that both techniques are the same, which is not true. But they do have some similarities in the way that they both try to capture process execution in reality and create process diagrams that help a business analyst to understand and analyze the process.

- **Process mining** is a technology that takes event logs from runtime systems—the "digital footprints" that we leave whenever we interact with systems—and creates diagrams from that timestamped interaction, such as creating a purchase order in the example used later in this book. 16

It is "passive" in the sense that it does not actively create records but rather takes existing information and visualizes it, as well as applying calculations to the information out-of-the-box.

- **Task mining** is different—it is an "active" recording of what users do, and depending on the tool that you use, it will do screen recordings, follow and record the document paths, etc. This is on the very lowest procedure level. Which field in this screen was filled? Which button was pressed? Which application was used before and after that step? Etc., and each individual activity does not make sense for itself, but the aggregation and abstraction of these procedural activities makes sense.
  To make task mining work, there needs to be special software installed on each user's machine so that it can record the activities.
- Please note that both technologies can only work with what is happening in IT systems. They cannot record activities in the real world, like filing papers or doing maintenance activities on an engine, but my hunch is that this will be just a matter of time when wearable devices, like glasses or pins, become more accepted.

So when should you use which technology? There are a couple of use cases where each technology has its advantages, but both can (and should) be used together:

- You can use both from a **hierarchical perspective**—you do process mining, and as one part of the analysis you find a step or process area that has anomalies standing out. Task mining can be a good tool to get more details on why step x takes longer than expected: do we have a user problem (the user doesn't know what to do and needs more training)? Is our user interface so badly designed that it requires rework so that the users can do the tasks more easily? Or do we have a process issue, and a specific vendor does not send the purchase order number on the invoice, and the user needs to call them and enter it manually into the finance system to create the correct relationship that allows them to pay the vendor invoice?

Task mining will not give you the answer on a silver platter, but the screenshots and other recordings will help to figure out the reasons.

- Task mining can also be used for **process steps that do not have a record in an event log**, for example, because a spreadsheet needs to be uploaded and that is not recorded as an event. Task mining can then determine which types of spreadsheets are used, how often, how long this took, etc., and this information can be added to the data model as an event log/data set for this process step and will show up in a process mining process just like any other step that was discovered from a system log.
- The additional information that task mining captures, such as where a user typically goes when they do activity 1 in system x, will **give context** that can help with the interpretation of the analysis data.
- Task mining can also help in the **discovery of processes that are not documented at all**, and when an organization does not even have an idea of what a napkin-level process diagram can look like. It can help to answer the "some finance magic happens here" question in the graphic below or even discover a complete process.

  This can be a serious accelerator for getting a first set of processes for an overall process management or audit initiative (and not only the analysis of a single process). However, be prepared that you will have to do normalization and aggregation activities that will lead to a manual modeling of a process diagram so that the result has the same level of abstraction and fits into your overall process structure of your BPM tool.

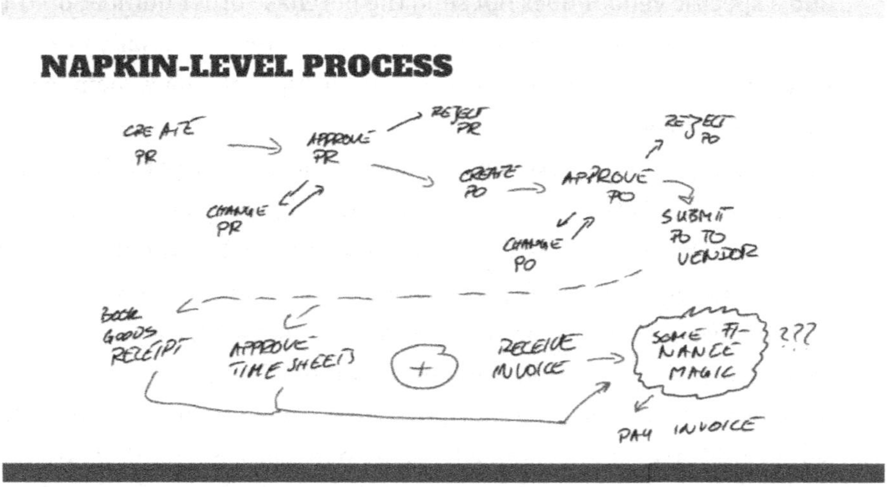

Fig. 7.1: Napkin-level process

There is one big problem with task mining, though: since it records everything, there is the "Big Brother" effect in place, which in the best case can lead to trust issues with the employees ("Why is my employer recording me?") and in the worst case can be plain illegal or create major headaches with works councils and unions for an employer. To any vendor's credit, there is functionality in the tools that will anonymize the user information, but there is still some residual trust issue. You need to decide for your organization how you want to address these issues when you think about implementing task mining.

For the purpose of this book, I will just focus on process mining, but be aware that in a more detailed analysis, it might make complete sense to integrate task mining as needed—it is "just another," separate implementation with its implications that would extend the scope of this book past its objective.

Case-oriented versus object-oriented process mining

The way to look at event data is different in these two approaches. The way almost all process mining tools work is to trace cases through a system. This has a few advantages and disadvantages:

Case IDs

The first one is a biggie: you must have a case ID. But what happens when your source system is not case-centric (aka has a unique ID for each instance of the process) but saves all events in a flat table (SAP, I am looking at you)?

In that case, you need to build the relationships in the data modeling phase following the activities first. In our example, SAP saves the events for purchase orders in the EKPO table. One of the fields in it is called BANFN, which represents the purchase requisition number in the EBAN table. Which means if your purchase order has a number filled in that field, you know that the steps "Create Purchase Requisition" and "Create Purchase Order" are from the same process instance/case. If the reference is empty, which can happen in the case of automated replenishment purchase orders, then you have a different case.

This can be tricky, and you need to understand the physical table relationships in a system, which might be a big ask for the average business analyst, so you might want to have your friends from IT close by.

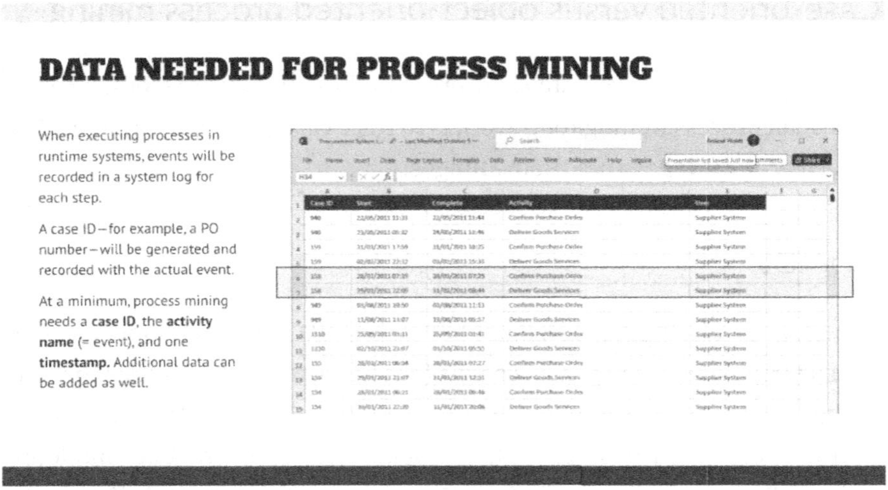

Fig. 7.2: Process cases in an export from a source system

Case-orientation

Another aspect is that you are following the cases based on a single item, like our purchase order. You will follow this main ID through the steps and can detect the time individual cases take if there is rework and the same case goes through the same step multiple times or via a loop, etc.

This allows you to trace variants and deviations from the "standard end-to-end process" (how work *should* be done) and allows you to calculate the time when cases sit idle between two steps—two of the forms of waste in Lean that you want to reduce. Being able to do this is a massive benefit and made process mining popular in the first place.

Object-centric process mining

But what happens when you have multiple items in a PO? How do you track these? The table in our example above just tracks the overall PO and the completion of the full activity.

This is when object orientation comes into play. With object-centric mining, you do not capture things in "flat" tables but rather in a database that tracks the objects (the individual PO items in our example) and the events that relate to them, as well as the object-to-object and object-to-event relationships. This allows you to create event logs "on the fly" for analysis purposes.

However, storing object information in a database format has additional benefits: you don't have to recreate the data model manually over and over again when you have the relationships already covered. The objects and events become system independent.

This allows you to reuse the same data in different contexts—for example, you can use invoice data in the procure-to-pay example that we have in this book, but you can also easily create a different invoice-centric analysis without the need to do steps 2 and 3 of our six-step process from scratch again (there is still some work to be done). Or you can combine multiple objects in one extended analysis and get more in-depth insights—for example, above, track the whole purchase order but also the individual items in it ... with the possibility to trigger interactions with the running process using standard alerting/notification functionalities that already exist in regular process mining tools.

What does that mean for me as a newbie?

Well, not as much as some vendors want you to believe. Process mining is not a tool-centric make-or-break thing; it rather is an assembly of analysis approaches that can be done with any process mining tool.

Do object-oriented mining tools give you more flexibility when you build a network of multiple, related process analyses? Yes. But it also comes with more complexity and needs more effort to maintain the database of objects and keep it current with each change in your source system environment. There is even a standard for this called OCEL (Object-Centric Event Log).[17]

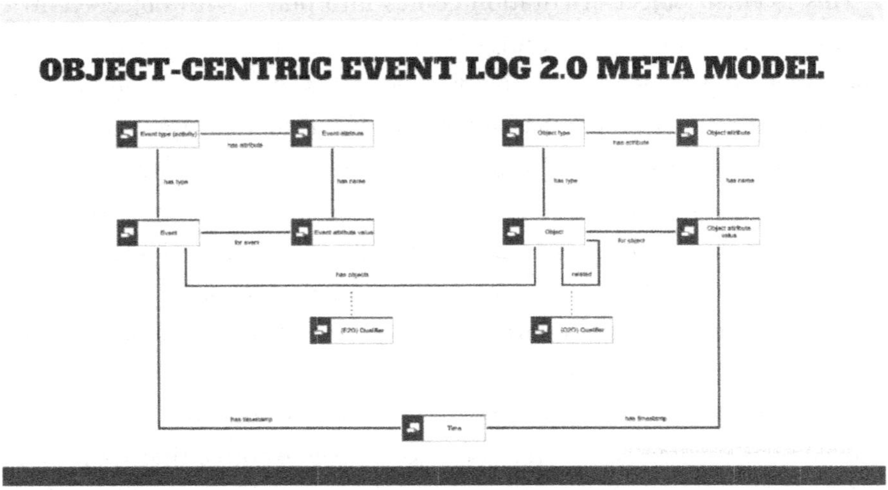

Fig. 7.3: Object-Centric Event Log (OCEL) 2.0 meta model

In the context of this book, we are looking at the easier, "traditional" process mining approach. This does not make our analysis "wrong," but it takes out the complexity of the data model and storage—tables are easier to grasp, and I want you to learn all steps outside the tool as well ... Simple is better, and you should "get the thinking" behind the approach and analysis :-)

8

Selecting a Process Mining Tool

Process mining is the "hot" topic in our industry these days, and there are many process mining tool vendors[18] available in this space. However, the technology is still at the beginning of the hype curve,[19] and not many organizations have established process mining as a standard tool in their process analysis toolkit.

In my opinion, it will take a few years for standardization, during which you will see some market consolidation. We've already seen some high-profile mergers and acquisitions in the process mining space, and I assume that in a few years you will have a smaller number of tool vendors to choose from. But until then, you will also see more innovation and the creation of ecosystems around a tool that will be beneficial for an organization looking for a "future-proof" solution. Most likely, you will see those coming from larger organizations that are not necessarily process mining pure players.

What is process mining, and how does it work?

Process mining is a relatively old concept, and early commercial tools were available as far back as 1998 (ARIS Process Performance Manager). But that was too early for organizations to accept into practice—they were busy implementing ERP systems and standardizing their processes for their chosen ERP system.

Over the last few years, there has been a renaissance in process mining tools, and a lively group of competitors has developed.

But how does process mining work? In a very simplified way, a process mining tool takes event logs from runtime systems and looks for steps that can be combined into cases, leveraging common identifiers as the common key (usually something like "order number").

The tool now takes the different instances and "discovers" the sequence of the steps, including additional data, such as frequencies, times, etc. When it has discovered different case types, it can combine the results and display a visualization of the process in the tool.

Disclaimer: The screenshots and conceptual graphics in this chapter show the ARIS process mining tool (aris-process-mining.com). However, all process mining tools work similarly.

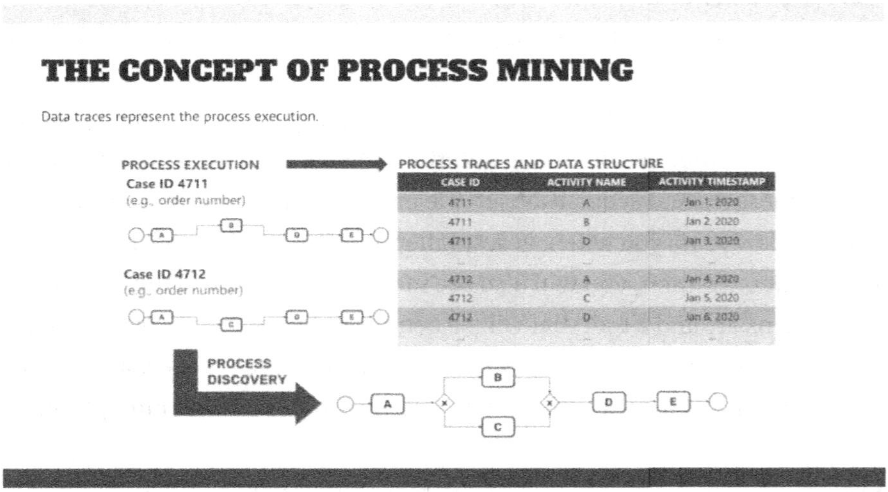

Fig. 8.1: Concept of process mining

What are features to look for?

Process mining allows for powerful analysis, and platform features provide a host of various out-of-the-box visualizations and configurable KPIs/metrics on dashboards and in reports.

But what should you look for in detail (as of early 2025)?

- The **Process Explorer** view (the graph visualization of the discovered process) is most likely what new users expect from a process mining tool. It shows all the paths that a process instance went through and displays various measures, such as frequencies (for cases and activities), times (for each step and the transition from step A to step B), or any other measure that you might have collected in your data set—things like cost or vendors, for example.
  This allows you to drill down to the finest details in a graphical interface, but it also comes with the need to develop skills on how to "read the data." For that, you should also have a basic understanding of statistics so that you don't get deceived by "an obvious observation" that is more an artifact of your metric selection (e.g., median vs. average in time calculation and the size of your selected sample vs. the overall case population).
  Keep in mind that causes for 'not-so-stellar process performance' are not one-dimensional, which means you want to have a combination of measures and drill-downs in mind when doing your analysis.
- **Variant analysis** is another crucial aspect for a process mining tool. As you can see above, it should be easy to select one or more variants so that you can see what is going on outside the "happy path." Ideally, you can export the discovered process variants into your architecture tool so that you can do further analysis and solution design for a future transition state implementation (see the Solution Lifecycle chapter for more details on the bigger picture).
- Once you find anomalies in your process execution and measurement, such as long processing times, the tool should support you in **root**

cause analysis with a root-cause miner. That feature automatically goes through the data set for this specific anomaly and allows you to see which aspect is causing the observation—does this happen with a specific vendor, for example, and can you detect a pattern? Or do things always happen at a specific location or with a specific user group?

- If you still cannot thoroughly understand the reason for the anomaly, you might want to take a look at **task mining** capabilities. Some tools have this built in, but in most cases, this is a feature of Robotic Process Automation (RPA) tools and might be named "Robotic Process Discovery" (RPD).

  This technology works differently than process mining by deploying robots that record screen activities. For example, a user just doesn't know where to click and needs more training, or information is missing (that vendor always forgets to add the PO reference), or a technical interface does not work correctly. The end users mitigate this by being the "swivel chair interface" and typing in data from another system manually.

  Another use case for task mining is the use of tools that don't even produce data, such as local desktop apps (like Excel). Task mining can capture these local activities and then provide an aggregated step as input for the process mining tool and complete the process.

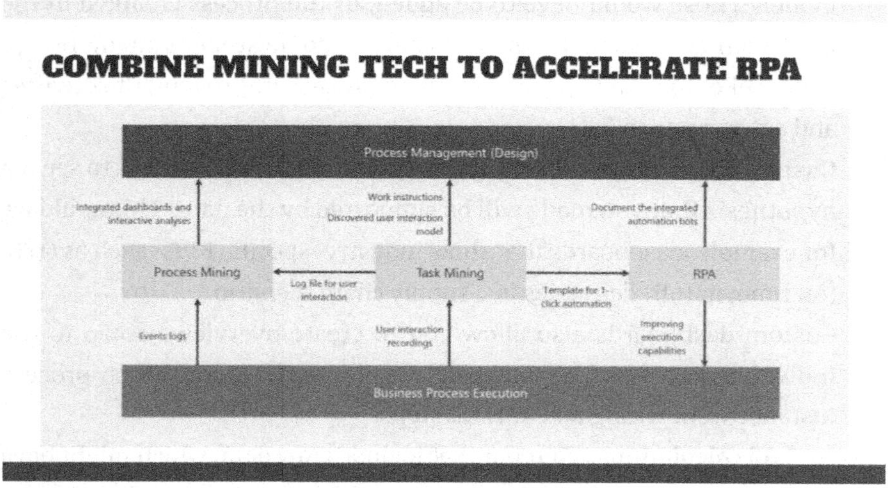

Fig. 8.2: Process mining—task mining—robotic process automation

Task mining, however, is a much broader topic that is outside the scope of this book. If you are keen to learn more about it, I recommend giving episode 9 of the podcast a listen.[20]

- **Conformance analysis** is another aspect you want to look at—to what degree do users execute the process as it was designed? A good tool allows you to upload a BPMN reference process and parse the data to see which instances/variants follow the reference process. Overview dashboards will show you the degree of compliance, either as an absolute number or over time. More advanced tools will also show you the "fitness value," which is the compliance calculation, not in a black and white view, but looking at the degree of compliance for each variant ("almost compliant except this" vs. "absolutely not compliant").
  The more practical aspect will be the detailed analysis of which conformance issues came up. Things like wrong start events (how many, how long did a case take when it started at the wrong step?), unplanned end events, mix-up of sequences (causing rework), or the identification of steps in the executed process that are not part of your reference process

model. Those would have to be added, or the process changed in the implementation so that there is no need to do these steps anymore.
A view like this can trigger immediate process improvement activities and act as a punch list.

- **Custom dashboards** will help you to investigate your data set to see if a hypothesis that you made will be supported by the data. This could be, for example, dashboards that show industry-specific KPIs, such as OTIF (on time, in full) deliveries in a supply chain scenario.
  Custom dashboards also allow you to create overviews down to the individual case level so that you can pinpoint exactly which process instance went wrong and at what step.
- **In-tool calculations**: Your data set includes raw data, which might have to be calculated to show, for example, if the segregation of duties has been observed. The tool should have a visual interface (no coding) for creating calculated columns that then can be used as a data source for a dashboard component.
  An example of this is "if PO creator is equal to PO approver, then separation of duty is not observed" and "if PO creator is not equal to PO approver, then separation of duty is observed." Both outcomes can then be displayed in a chart component on the dashboard.
  Since these types of data model adaptations are done by business analysts, a graphical user interface for creating the if/then/else statements, with options to nest these, is crucial. In the rare case that you will have to enter some query language, the better tools will provide some AI support so that you can enter what you want in natural language and the tool creates the code it needs.
- Speaking of AI: tools should have some form of **prediction** capability to determine when a case will end and which path it will most likely take, using machine learning. This can be used to identify the cases for which you might want to take action immediately and not wait until the (negative) event has taken place.
- **Data transformation** is the activity that takes the raw source tables and transforms the fields into the format required for the process mining

tool. Typically, this is done via SQL-like statements ("join," "rename," and so on). Since this is an "engineer" type of role, whoever does this will find basic coding skills and data modeling skills are helpful for doing this activity. Or you follow the advice here and use a visual data modeling tool and simply don't use the built-in transformations when preparing your data model.

- **Integrations**—this is a larger topic, and the next section will cover the four integration types in more detail.
- **Export and import of "solutions"**—prepackaged analysis that already exists (for example, provided by consulting organizations) and can be imported easily and will give you pre-built dashboards and instructions/ templates for the data collection.

Integrations

By now, it is clear that your process mining tool must integrate with other systems. However, there are four different aspects you should look at when evaluating integrations.

Data import from runtime systems

Typically, the default import of execution data is via a CSV file. This obviously comes with advantages and disadvantages—the advantage is that you don't have to build an integration with the source system, and you can rely on reports from the source system. The disadvantage is that the data is not "live," and you might have to transform the source data outside the process mining tool into the format that the process mining tool can process (see above).

You might want to consider a live integration, and for this, you typically see two options:

- Dedicated **extraction adapters** for source systems or generic adapters, such as a JDBC adapter, that can be used on any database supporting this protocol. For obvious reasons, this approach has its limitations (there are too many potential source systems out there), but you should look for support for major business applications.

- The use of an **integration tool**. Integration tools typically have more adapters available (in the hundreds or even more), and they also allow the development of custom interfaces. While this might help connect a lot more systems, it also comes with more complexity and the potential need for coding skills.

In the graphic below, all three import scenarios are depicted using Software AG's tool stack.

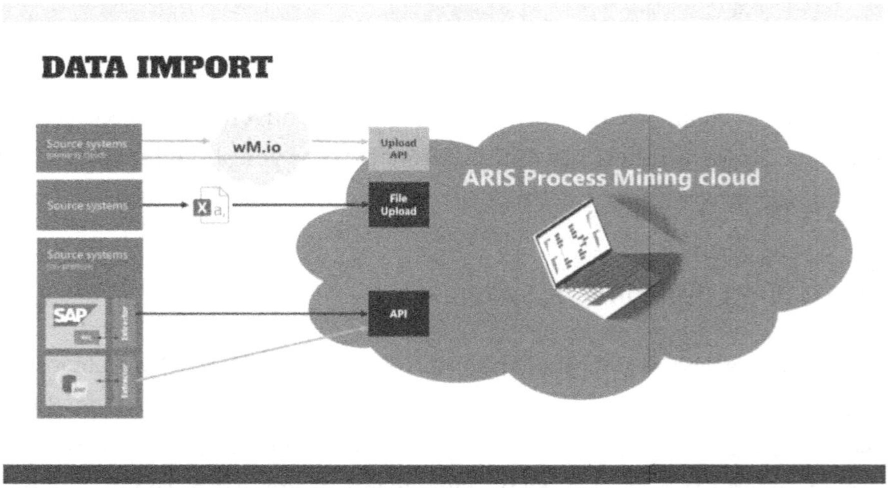

Fig. 8.3: Example of schematic integrations for ARIS Process Mining

Integration with a task mining tool

Some process mining tools have task mining capabilities built in already. In case your chosen tool does not have this, you should look for a possibility to integrate task mining results into your process analysis (see some example use cases above).

Integration with your architecture tool

Process mining is an excellent tool to get a profound understanding of your as-is processes and, depending on your data set, related aspects such as cost or compliance. However, it is not meant to create a future state design, and it is limited to the underlying data set.

For these aspects, you want a proper architecture tool that allows you to do analysis or simulation, even if the future state process is entirely different from the discovered as-is. In addition, your architecture tool should support the full solution lifecycle, which means it needs to be capable of containing your enterprise baseline to resolve the situation where "what was designed" does not match "what was implemented" (the potential discrepancy between the solution design and implementation phases), for example.

Therefore, you want two integrations:

- **Upload of reference processes** from your design platform into the process mining tool (see above).
- **Download of discovered processes** from process mining to the design platform. This feature allows you to skip as-is workshops with stakeholders and SMEs and will produce a baseline model that is based on the actual process execution.

An added benefit of the second feature is that you can use a "real" simulation engine that allows you to set parameters for your future-state process that might be thoroughly different from your as-is data.

Using process mining analysis results to trigger actions

The last integration that you are looking for is the ability to trigger actions based on process mining results, such as process instances or steps that run too long. You should be able to define the conditions that will trigger the action—times, volumes, values—and then provide a web hook or native integration with a workflow component. This could be either built into the process mining tool or trigger an integration engine that you might already have in place (in the graphic above, it would be the webMethods.io tool). Potential actions could be notifications and alerting or the execution of transactions in runtime systems.

With this integration, the lines between a design/analysis tool and a runtime tool start to blur, and you should take a closer look at how deep you want your architecture group to go down the rabbit hole. Since this can be quite an effort, my recommendation is that you work with the business and IT groups that are responsible for the runtime tools and stop at providing the hook for the workflow component but not operating it.

Hosting and cost

Most process mining tools are now "cloud only," and this seems to be the world that we are living in. Some vendors offer on-premise versions of their tools, but have a close look to see if these are the same offerings that are available as SaaS versions or are older legacy tools that most likely won't get updates anymore.

In addition to this, you will also have to look at the applicable regulations and how your vendor handles these. Examples that come to mind are things like GDPR (the European General Data Protection Regulation) or the SOC 2 standard (System and Organization Controls),[21] but I am sure that your IT department already has a long list of requirements for any hosted tool that you and your vendor will have to satisfy.

If your process mining tool uses AI, you might want to check where your data is stored and if it is used to train the AI so that the vendor will then use your information to make it "better" for others—something that you might not want.

When it comes to licensing, there is no "standard" pricing between vendors, but the components of the price tag are similar:

- A **base fee** for providing the process mining tool. This typically includes the CSV upload and some base storage and/or number of cases. Please note that typically this fee will be calculated for one (1) process that shall be mined, independent of the number of applications you pull data from. That means that you should take a look at your end-to-end process structure and see how many of those will be candidates for process mining. Since the overall price tag can grow quickly, it is worth revisiting this.
- **Storage or number of cases**—those numbers can start from 250,000 cases being included in the base fee, up to hundreds of millions for a "4XL" package.
- **Data extraction**: depending on the type of integration (specific extractors or [cloud] integration tools with more flexibility), consumption fees might occur. While the extractors might be included for free, cloud integration components typically come with a consumption price tag (number of cases, most likely).
- **User subscription fees** are the last category of costs that you might encounter. Typically, you see relatively cheap viewer roles, business analyst roles, and technical engineer roles being defined that might be needed for the different types of integrations vs. just the base tool.
- **Solution costs**—as discussed above, some process mining tools allow prepackaged analysis for instant use. Those might come with a price tag if they are not included in the base tool and will be provided by consultants or other members in the tool ecosystem.

It is worth the time to take a close look at what your needs and data volume are, as well as what the package from the vendor might look like at the beginning of your negotiations. Keep in mind that you most likely will have additional consulting costs, especially when you are new to this whole topic and you don't have an infrastructure established (and now take a deep breath and get familiar with an initial price tag of $40k-100k for your first process) ;-).

9

Buying a Process Mining Tool

My assumption is that most readers of this book are working in a "regular" organization and are looking for a tool that will help them to do their work. They are not procurement specialists or working for a vendor.

This chapter will look at the process of buying an enterprise tool and also look at the "other side" of that process—the vendor's perspective. This might give you insight into why they react how they react to your questions and demands.

Who is involved in the buying process?

There are different players involved in this process—on the buyer side and on the vendor side. Let's take a look at them:

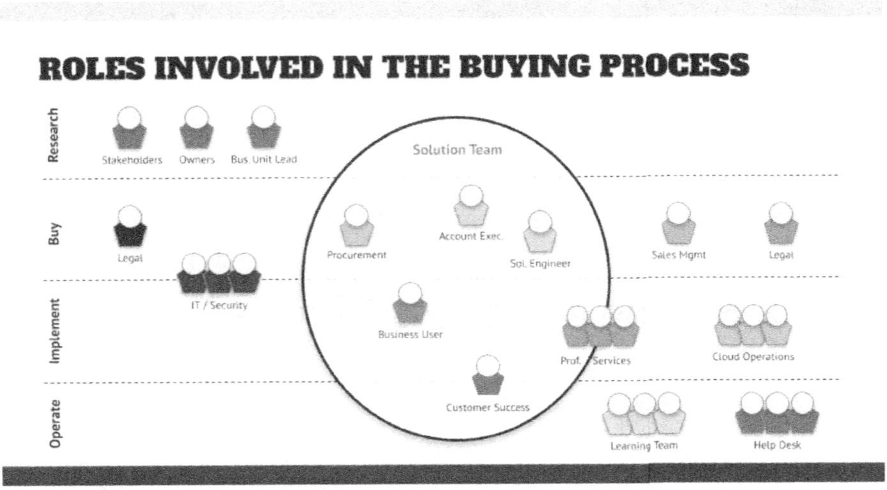

Fig. 9.1: Roles involved in the buying (and implementation) process

- The first one is the **business user**. This can be—theoretically—anyone in the organization who has a problem to solve. Most likely, you will be part of a BPM group or work on a project that needs to get a solution designed or improvements identified.
- If your organization is a bit more mature, you already might have some **owners** defined who are either responsible for an area (like process owners or application owners). Or you have set up whatever you could set up in an enterprise-wide organization, and you run out of "juice" with your office files, and the management of the artifacts becomes too much so that you can't find what you need for your process mining efforts.
- I also have seen **business unit leaders** who jumped on the process improvement/optimization bandwagon and are looking for a tool that they can "grow into." Occasionally, these folks are not hands-on involved in the buying process but need to get educated so that they can approve the expense and provide their sponsorship and support to the project.
- **Purchasing people**. The pros in your organization spend money and

negotiate the best deal for products and services. They can be a pain at times and be "overly formal," but be assured that they know how the game is played, and you want them on your side.

- Your **IT people**, who already might have people who call themselves architects or who need to approve the inclusion of new software in the company's ecosystem. You need these folks because you need their operations support and resources to have your tool run reliably and securely, especially when production data might leave the organization because your process mining tool runs in the cloud. There might be requirements from the IT security people that your chosen tool follow certain standards or make it mandatory that you anonymize the data before it leaves the premises.
  This group also includes the people in your organization who manage IT security.
- **Other stakeholders** like risk and compliance people who might have specific requirements for the software that must be included.

On the vendor side, you typically see these roles:

- The **Account Executive**. This is the salesperson who is your go-to contact during the buying process. These people manage the initial scoping conversations—and the better ones will give you initial demos—and will assemble the vendor team for your request.
- **Solution Engineers**. These are the rock stars in the process, and you want to build a good relationship with them. They listen to your requirements, translate that into solutions built on the software, and are also the knowledgeable tech people who will educate you and your colleagues on the abilities of the tools and the concepts behind them. Solution engineers also build the demo and proof systems and will accompany you during proof of concepts, etc.
  And the better ones will also tell you when their software is not the right thing to solve your problem.
- **Sales management** are the bosses of the previous two roles, and they get

engaged when it comes to non-standard sales situations. Maybe because you are a large organization with special requirements, or you dangle a massive carrot in front of the salesman (remember: most of them are "coin operated"). These roles on the director or VP level will also be the ones who will make the case when it comes to discounts, etc.

Sometimes larger organizations have this bundled in a (global) deal desk ... the counterpart to the procurement organization in the buying organization.

- **Legal**. In both organizations, there will be lawyers who want to give their stamp of approval when it comes to contracts that are not standard. These folks are also involved in the beginning when it comes to non-disclosure agreements (NDAs) so that you can even start sharing information.
- **Professional Services**, or a **partner** organization, who will provide the offer for the implementation and other services that you might request.
- **Customer Success Managers** (CSM). If your deal is over a certain size or non-standard, you might get a CSM assigned. They will take over as your day-to-day contact after the sale is done, and you are either in the implementation or, thereafter, in the production phase. CSMs will work with you on success plans and share their knowledge from other implementations and, in many organizations, are also responsible for renewals.
- **Learning organizations**—the better vendors have dedicated groups for learning and enablement. These typically offer a wide range of education products and services that can span from self-paced e-learning through instructor-led training to formal certifications. They also might maintain a knowledge base for customers, but typically, you find the learning specialists in these organizations, and they might be able to answer "standard questions." If you have specific questions, the professional service people and the solution engineers might be the better contacts. After the sale, the CSM will be your go-to person.
- **Cloud Operations** is the group that will speak with your IT security folks and provision the software and do some basic configuration like the single-sign-on setup.

- **Help Desk**. Typically, vendor organizations have a three-tier help desk setup: a first level might have dedicated individuals who are available at certain times and take on your questions and might be able to help you with them. Depending on your vendor's size, this can go from "business hours Mon-Fri" to "24/7/365" with help desk people in different locations and time zones.
  If the first level can't answer your questions directly (the good firms have built knowledge bases that they, or even you, can access), your request will be forwarded to dedicated second-level people, who are specialists in the different areas of the software. If the problem goes deeper, the third level typically is the product management/development group, and you might have found a bug that needs either an immediate fix (in case of security issues) or to go on the planning board/roadmap for future releases.
  In either case, the first-level help desk will be the main communication partner for you and manage your requests via tickets (and in some organizations you can look up the status of your ticket by yourself).

The buying process (and thereafter)

Buying a process mining tool is typically an enterprise software process that we will discuss below. There might be some vendors who offer self-service without engagement of humans, and that might be enough for an individual project. Typically, you will not get support past the learning offerings and are on your own to learn via the documentation.

The process for buying an enterprise-level tool is shown in figure 9.2 below:

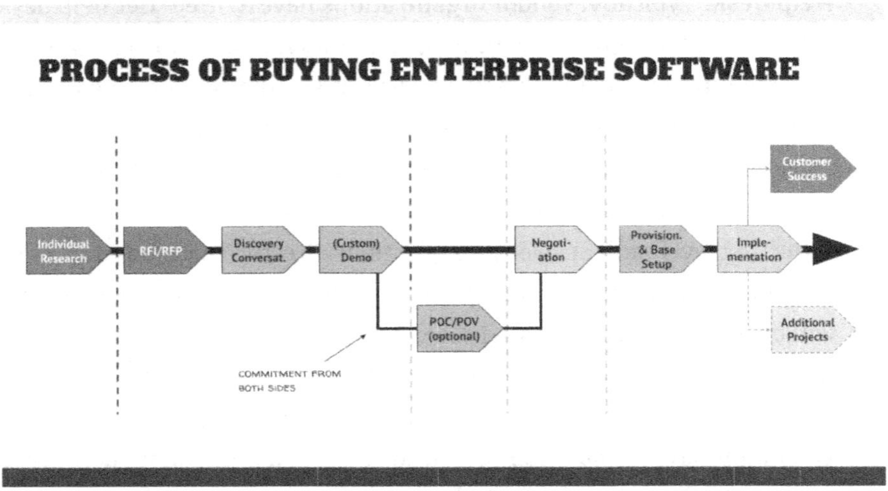

Fig. 9.2: Process of buying enterprise software

Internal Research

The process starts with the business user doing his or her research and self-education on use cases, features, and other specifications, such as tiered offerings, etc.

This phase typically is very unstructured, and people come up with a list of things (like "must provide this type of analysis," "must support notation XYZ for interchange with architecture tool," and "must integrate with existing system ABC") and an initial market list.

However, only a few organizations that I have worked with have gone the extra step and already thought about the different personas/roles that we will speak about in a different chapter below. They have then assembled what life with the new tool will look like for each persona.

You also want to include the other stakeholders from above in this early research so that you learn about their needs and requirements, especially regarding security, cost, or budget.

One thing that Google has researched is that it takes 11 interactions with a brand and 7 hours of learning about the tool and the vendor before customers even attempt to reach out to a vendor. So, vendors, you better have good content on your website and make material available in the form of podcasts, webinar replays, marketing material, implementation guides, etc.

As indicated by the first vertical dotted line, this might be the first point in the process where a potential buyer checks out of the process and does not include your tool on the list for the next steps. And a vendor might never learn about this opportunity.

RFI and RFP

These acronyms stand for "Request for Information" and "Request for Proposal" and basically mean the same except that an RFP would like to see specific estimates for all aspects of an implementation (in dollars), while the RFI might only include high-level "ballpark" numbers. The RFI can be issued at any point, while the RFP typically comes a bit later, when the business users have made a pre-selection of vendors and the requirements are more solid.

Typically, these docs will be issued by a procurement organization, and there has been an internal process of purchase request done in the buying organization, including approvals by leadership. In the majority of cases, a buyer organization sends over instructions and a big Excel file with the requirements listed by functional, non-functional, security, etc. requirements. There also might be a form to be filled out that will hold the pricing information. The expectation is that a vendor will fill out these documents, but in almost all cases I advise against that.

Discovery conversation(s)

The better way is to have a conversation with the salespeople ("discovery conversation") in which you give the vendor a wider overview of what you want to accomplish. This way, the account executive can assess if their tool is a fit on a higher level and which team members need to be assembled.

Then a second call (and maybe additional calls) that includes a solution engineer should be scheduled. The objective is that this is not just a checkbox exercise. You have open conversations and define a solution that includes the technology and takes all other aspects of an implementation—which we will discuss further below—into consideration.

In my experience, this phase needs deeper conversations, and words will be interpreted differently. The ideal outcome is that there is a shared understanding (ideally in writing, which includes diagrams) of the solution that will be implemented. This includes everything from the objectives through the technology, implementation, and measurements on a high level.

(Custom) Demonstration

This is a phase that might run in parallel to the discovery and solution-shaping conversations.

It begins with a standard demonstration. Most vendors do have their use cases implemented in their tools using dummy data so that they have a context in which the artifacts will be demonstrated. Depending on what you are shopping for, for example, a modeling tool or a simulation tool, which you might need besides the process mining tool as shown before, that might be enough to give you an impression of the tool to be ready to score it.

Some organizations have a more peculiar stakeholder group and want to see their content in the vendor's tool ("custom data"). This should not be a big issue, since all tools will have some form of import, and the artifacts are created relatively quickly. The exception to this is process and task mining, which require an example analysis and will follow the steps of a "live" mining project, which is beyond the scope of a demonstration.

This phase will be done by the solution engineer and supported by the account executive, who will provide available additional information in the form of brochures, white papers, webinar recordings, and so on.

POC/POV

POC and POV stand for Proof of Concept and Proof of Value. The difference between the two is that the first has a more technical focus ("Does this thing work for us?") and the latter has a more content focus ("Does it do what we want to do with it?").

Proofs are an optional step and at this point should only be triggered if there is a serious interest from a buyer. It will involve a time commitment and engagement from both sides and multiple players and will be run like a real project with defined outcomes and deliverables—but on a smaller scale. The underlying understanding is if the tool fulfills the objectives of the proof, then the buyer will proceed to the next phase, the negotiations.

In a proof, you will see not only professional project management, but they also typically include enablement activities and giving customers access to a separate technical environment that is provided by the vendor. If the content is sensitive data, there even might be an installation on the customer's premises. However, the main idea is that the buying team gets their hands on the tool while being guided by experts from the vendor so that no misconceptions occur. A good vendor will also tell you if your requirements are too high, the tool can't do things, or how issues will be addressed in future releases of the tool.

Depending on the effort of the Proof project, be prepared that the vendor wants you to pay for his efforts. This also could include the deduction of the proof cost if a customer buys the software, depending on the negotiation skills in the next step.

Negotiations

This phase starts with the decision on the buyer's side of which tool they want (or the shortlist of not more than 2-3 tools) and might start with a formal RFP to be filled out, which might be just a formality at this point.

However, the vendor now will submit a formal offer and all the papers that need to be signed, and you will go into a negotiation phase. There are multiple levels at which the various salespeople can give you discounts (for example, if you subscribe for a certain amount of time and not month-by-month, or if you have a progression in the number of users or cases in the case of process mining), but there are limits to this, and your account executive might have to chase multiple levels to come to an agreement that is acceptable for everyone.[22]

Depending on the type of contract (for example, did you buy a pre-packaged set of tools and seats, or do you have a custom enterprise setup?), the lawyers of both firms are involved, and they might have their special ideas to be included. One thing that I recommend negotiating in the first package, though, is the Master Services Agreement (MSA) that will rule all services that will be provided to you. In the future, you just have to agree on Statements of Work (SOW) for project phases without having to agree on payment terms, processes, etc. every time.

Provisioning and base setup

Congratulations! You bought your tool. Now the real work begins, and the first step is that you get access to your instance (assuming that you are using a Software-as-a-Service/SaaS) provisioning. Otherwise, the installation and base setup are part of an implementation).

This typically happens when the account executive triggers the vendor's cloud team to send the credentials for a superuser to the named customer person. That team will then start the conversations about the base setup, which typically includes the configuration of single sign-on or of an email server so that the tool can send out notifications. Depending on your organization's (security) requirements, there might be additional technical things to be solved.

What you don't get in this step is detailed training and the creation of content/structures or users in your system. These are things that happen in the implementation phase.

Implementation

The implementation phase is typically done with the help of the vendor's professional services organization (PS) or one of their partners, and you have been introduced to them in the sales phases before. If they are good professionals, they will follow the approach that I describe in this book and that is presented in an overview in the "Process Mining Implementation" chapter above ;-)

At this time, the sales team will disappear, and the project manager will be your main contact. However, good account executives will keep in contact with you and check in occasionally so that you don't have a "cold" request if there is the need for more licenses.

The not-so-good account executives will see themselves as the single point of contact in their organization and demand that any request always go through them. Unfortunately, many topics in the implementation or later customer success phases are over their head, and they don't bring a lot of value to the conversation, or rather, they are delaying the solution of problems with their behavior. Make clear that you understand their role and that you'd rather work with the correct people in their organization directly.

Working with the services organization of a vendor/partner is a good thing because these people have seen basically all questions that you might have and are involved or are part of the vendor's learning team. However, the challenge is that professional service organizations normally work on a project basis, and when your project ends (successfully!), they move on, and your contacts might be a dedicated customer success manager or the help desk.

And, of course, the professional services organization will be happy to have additional projects with you. Those will be directly scoped and negotiated with the PS organization, and vendors do have different processes for service and license deals. As part of that, use the experienced PS resources that work with you to develop and update your roadmap and grab their knowledge of what has worked for other clients who have been in your shoes before.

Customer Success

If you are a larger customer or have a strategic importance for the vendor, you might get a Customer Success Manager (CSM) assigned. Typically, this person will be introduced during the negotiations phase so that they will observe what is going on in your program "as it happens."

The main objective of the CSM organization is to make you successful in reaching your business objectives that you have defined for your program. As part of this, they will schedule regular meetings with you (typically on a monthly basis) and develop a client success plan together with you, which will also include vendor resources where applicable.

The CSM will be your main counterpart on the vendor side for the duration of the contract that you have closed, and I highly suggest building a good relationship with them. They can be the necessary "accelerator" in the vendor organization, which might be needed, and they can arrange meetings with the right resources and individuals to solve a problem, even if this is not "part of the standard process"—something that you want.

This also means that you should keep an overview of all open help desk tickets that you might have and keep the CSM current on the progress (or lack thereof) of your program or the initial implementation.

Lastly, the CSMs in some organizations are also responsible for renewals when the contract period comes to an end. Make sure that you get the value from your investment and discuss your license (and services) needs with your CSM in advance so that this does not become a "finance organization sends out an invoice" exercise, but both parties can track the completion of the customer success plan and have a good understanding if the assumptions that were made during the negotiations, such as the actual usage of the tool, are fulfilled or need to be adjusted.

Final thoughts

As you have seen, there are many people involved in the process, and all of them operate under different conditions and have different, sometimes conflicting objectives. To avoid any problems or to de-escalate conflicts, I recommend "playing with open cards" once you have signed a contract.

In addition to this, you should clearly communicate what your expectations are, hold your end of any commitments, and be a reliable partner through the sales, implementation, and customer success phases of your program. To get a head start on this, I recommend developing an initial strategy as described below and having clearly defined personas before you even start the interaction with a vendor—clarity will be appreciated by everyone who is involved.

10

Process Mining Implementation

Implementing a process mining tool for a limited scope can be an easy exercise because it is "just" another tool set for analysis purposes. And when you look back into the first step of our six-step approach for successful process mining projects, we have already covered some aspects:

- Strategy for Process Mining
- Stakeholders/SMEs and their involvement
- Planning for this capability in your organization
- Identifying the right process in scope and the desired outcomes

However, typically, you will need more tools than just a process mining tool, and we will talk about that in a later chapter in more detail.[23] These tools, like a process design or process simulation tool, are most likely already in your organization, and you "just" add one more tool to the stack, which might be as simple as distributing licenses and making training accessible.

However, in case you don't have this larger setup, there are more things to do.

Implementing the larger tool stack

I, literally, have written a whole book on how to implement an architecture tool stack successfully, so I will just give a short overview of the four areas that you might want to look at here. If you are interested in more details, you might want to look into 'Successful Architecture Implementation'—I appreciate it.[24]

The approach to bringing in architecture tools is to treat the program as an enterprise application implementation that is driven by the introduction of a new capability for the organization. This is obviously a larger effort that needs to be led by the business and not by IT. Doing it the right way will lay the foundation for process improvements, system implementations, business/digital transformations, and regulatory compliance.

The full approach for implementing an architecture capability, which includes process analysis, includes four areas, shown in figure 10.1 below.

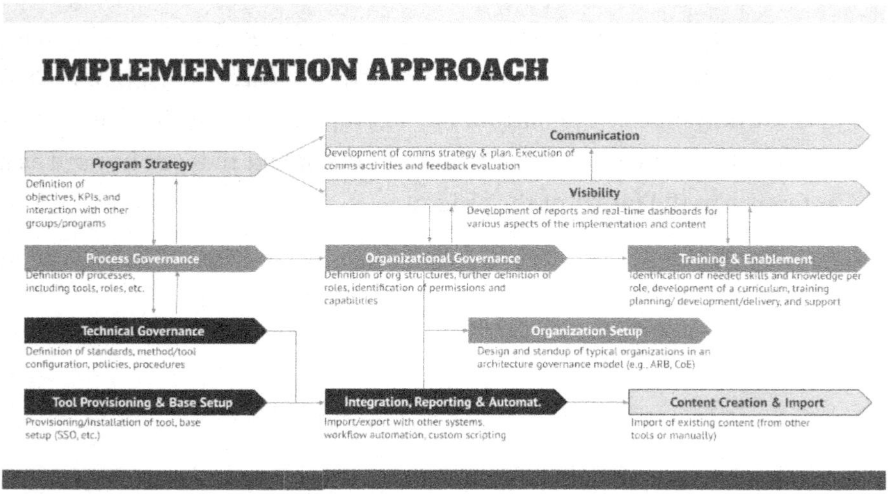

Fig. 10.1: Architecture implementation approach

The architecture implementation approach covers four areas:

- Strategy
- Governance
- Technology
- Content

Strategy Work Area

This work stream (light green in the graphic) aims to find the proper place of the program within other initiatives and manage it.

- The **program strategy** phase looks at the functional scope of the program (e.g., visualization of processes and where risk and controls will ensure regular compliance on the 'who-does-what-when-where' level, typically depicted in BPMN diagrams) and identifies the 'sources of truth' for information, as well as where the architecture tool becomes the main source of information. This will then also determine how the program will interact with other programs and/or business-as-usual org units that are responsible for systems, as well as the necessary IT/development support (and standards) that are needed to create integrations. This is shown in blue at the bottom of the graphic. Obviously, this will be captured in a "living" program strategy document that can be used as a reference in the future of the program.

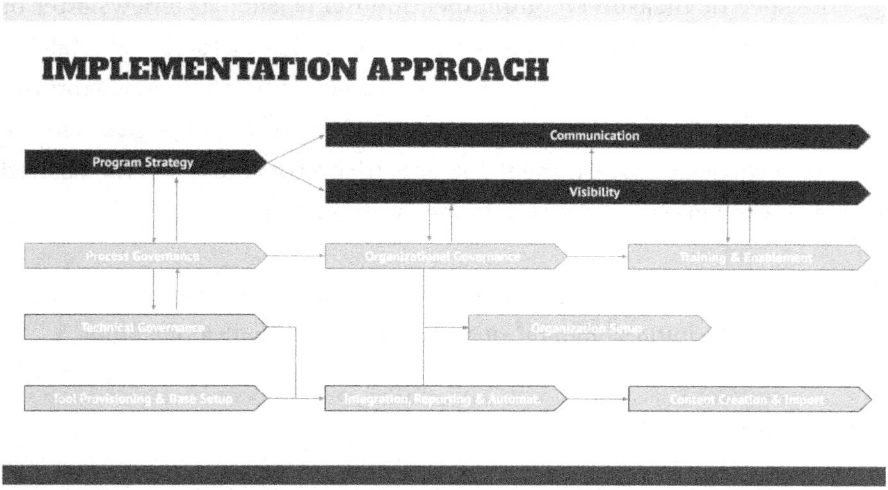

Fig. 10.2: Strategy work area

- **Visibility**—The next step then is to define and formally describe the objectives of the program over multiple releases, as well as the KPIs to measure success, and identify where they come from. This, and the identification of stakeholder information needs, sets the foundation for the development of reports and real-time dashboards in the Visibility phase that support the management of the program. All artifacts will be done in a solution architecture database, which will be extended in other work packages and serve as the 'meta' information source that will be used when interacting with other entities.
- **Communication**—to be successful in your implementation, you need to consider these three areas: content, governance, and adoption. One of the big things in the adoption area is to start communicating as early as your program was approved. To do this successfully, you might want to start laying out the foundations for it while developing the strategy in the form of a communication strategy, which covers stakeholder groups, key messages, communication and feedback channels, and so on. Based on that, you can build a (repeatable) communication plan that will include

the specific measures/communication items, such as a newsletter or event, and the responsible people to execute it. The last step is to establish and maintain feedback and support channels, for example, via forums, regular feedback collection meetings, and so on. The objective is to use the feedback to determine if you are still on track to reach the desired business capabilities laid out in the program strategy.

At the end of this phase, the foundation for stakeholder reporting and visibility is in place, and the initial communication has started to create momentum.

Governance Work Area

A second 'housekeeping' item is to define the governance that applies to the program (dark green in the graphic). It typically is set in four work packages that interact with other work packages, and all designs are developed in a solution architecture database.

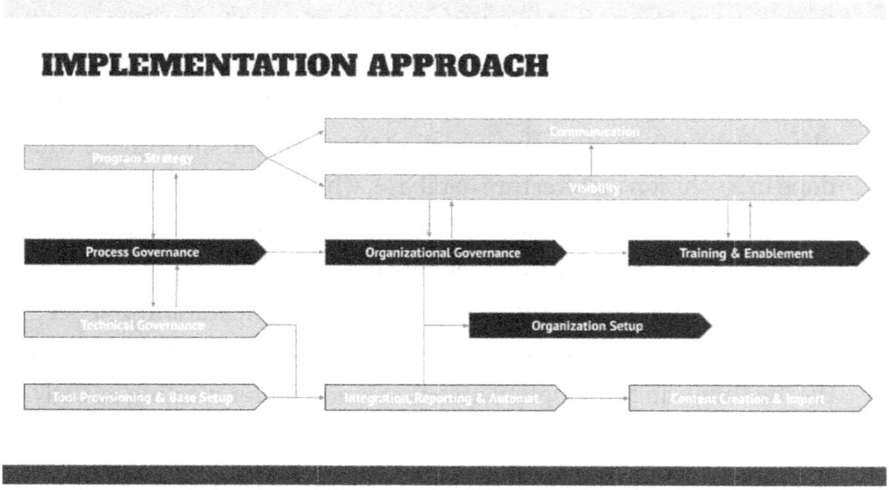

Fig. 10.3: Governance work area

- **Process Governance**: The definition and documentation of the processes (including "auxiliary information," such as to-be-created artifacts, used applications, etc.) that users, workflow participants, admins, and stakeholders do. These cover processes such as the creation or import of models, reviews and approvals, or the management and operations of technical interfaces that will be created with other systems. In addition to these policies and compliance requirements, such as the level of detail or the frequency of architecture artifact reviews/certifications, they will be developed in this step. Typically, the models created in this phase consist of one high-level Value Chain Diagram with assigned BPMN processes underneath, which will include the business steps, roles, applications, etc.—the 'who does what when where' level processes of the program.
- **Organizational Governance**: Based on the defined processes, roles are identified, and permissions and privileges will be defined and administered in the tools. The main artifact for this is a license/permission/privilege matrix that includes every role. This work package also defines the organizational structure that will run the tools in production, such as a Center of Excellence (COE), a standards group, or an Architecture Review Board (ARB).
- **Training & Enablement**: Based on the process and role definition, the skills and enablement needs will be determined, and a curriculum will be built for each role. This not only includes the formal training but also additional support measures. Based on this design, the training and support material will be built and delivered.
- **Organization Setup**: In this work package, the identified org units will be set up, candidates identified and selected, and the initial meetings supported by the program management until the org units are self-sufficient in their operations and planning/management.

At the end of this phase, the "rules of engagement" are defined, the needed org units/teams are set up, and users are trained in the use and administration of all systems used in the program.

Technology Work Area

This work stream (dark blue in the graphic) is typically the only one that clients think about when it comes to the implementation of an architecture tool.

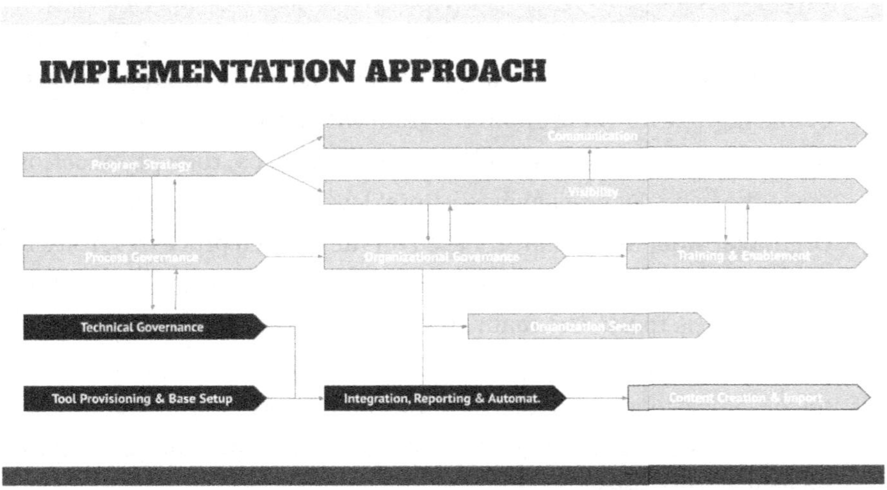

Fig. 10.4: Technology work area

It is typically divided into three work packages:

- **Tool Provisioning & Base Setup** aims to make the tools available to the program. Depending on the hosting (cloud or on-premises), this will include different steps and participants, but typical activities include the setup of single sign-on, connections to the network, integration with email servers for notifications, etc. At the end of this activity, the tool vendor hands over the keys to the tool, and the program takes over responsibility for the tool. While the vendor and/or the IT organization manages the technical aspects of the tool, such as backups, user group creation, etc.

- **Technical governance** determines the configuration of the tool and the underlying method decisions. This includes the creation of data models and interface matrices (to other systems), as well as a metamodel that shows the levels and views, as well as the chosen artifacts at their intersections. These artifacts (typically models) are defined by their notations, objects, attributes, etc. This work package also includes the database setup and structures within each database. The data models typically are added to the solution architecture database, while the metamodel is created in a configuration database. This allows the creation of the tool configuration files (filter and template in the case of ARIS) automatically by running a wizard on the configuration DB, which saves the time of the manual configuration of each item.

It also facilitates the workshops for the data model and metamodel design and gives stakeholders a first glimpse of how the new tool will be used in other solutions.

- The last sub-work package of **Integration, Reporting & Automation** picks up the deliverables of the Strategy and Governance phases and includes the creation of scripts for technical integration purposes, visibility reports, and the automation of (governance) workflows within the architecture tool stack.

While doing this, leading IT practices, such as a defined development approach or the use of multiple environments for development, testing, and production, should be considered. Stay away from implementation partners who see this development as 'one-offs' that do not have to follow these practices.

At the end of this phase, all tools in the program are stood up, configured, and ready to use.

Content Work Area

The last step in the implementation (light blue in the graphic) is the creation and import of existing content and/or reference content from standards organizations, such as APQC[25] into the content (work) database.[26]

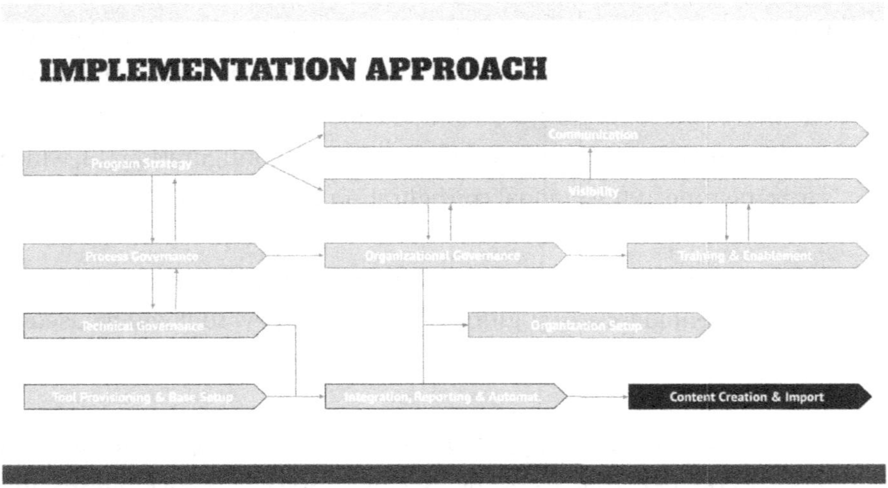

Fig. 10.5: Content creation and import work area

Existing content might exist in various formats—Excel, Visio, *.bpmn files, etc.—and need to be imported into the tool. This can be a larger effort, since the main work is not the initial technical import (this is typically done well by the tools) but the normalization and integration of the content into existing hierarchies.

Another part of this work stream is the import of reference data from upstream systems, using the interface scripts from the technology work package. This data typically comes into the tool as catalogs, and they might have to be rationalized. Other imported content, such as processes, needs to be reviewed to see if it used the correct content from the system of truth. This is in case there are discrepancies or needs for consolidation of objects to ensure consistency of the database and the content structure.

At the end of this phase, the architecture tool is ready for a rollout to a wider audience.

What does that mean in the context of process mining?

Process mining can be run as an isolated project (the focus of this book), but most likely is embedded in a larger business process management program, which has these things hopefully already figured out.

In case that this is not the case, you should consider the following things:

- Define a clear goal and objectives for your process mining project. Is this a "one-off" to test if you want to implement this capability, or do you want to change your approach to process analysis towards a data-driven methodology?
- Define how you would like to do process mining in your organization. Which org unit will start? Which roles do you need for this initial project? How do you scale and prep the next org units?
- Define the organizational setup for this capability. Do you want a central support unit, and what does this unit do, or do you just want to manage the tool and hand out licenses, and the other org units are on their own (an approach that I do not recommend)?
- Aim for short turnaround projects—four to five weeks in a simple setup. You want to build momentum here and not, as you would do in establishing your architecture repository, create a structure that will last for years to come.
- Having said that, your architecture repository and its content and other tools in the stack will be needed to do a complete process analysis with your process mining tool. For example, you will need a process that describes how work should be done so that you can compare it with your actual data set and find improvement opportunities.
- Training and enablement is the next topic you want to think of. The vendors typically provide tool training and a bit of methodology sprinkled

into the training. What you would like to add to this are these topics: basics of statistics, data modeling (concepts), process analysis concepts (for example, Lean, Six Sigma, etc.), and BPMN notation as the standard in describing processes.

If you have business process management already implemented, I am sure that you already have material about these topics, but I want to convey the message that you need to know more than just which button to press—process mining is not a "load the data and relax" approach, as you will see in the next chapters.

- Having said that, the technology aspect is typically pretty simple. Almost all process mining tools are hosted in the cloud, and you don't have to worry about installation, maintenance, etc. What you want to check with your IT and security folks, though, is what rules for data upload or AI usage you have in your organization, and have them check the vendors so that you don't unintentionally "publish" internal data.
- Lastly, you might want to think about reporting and developing standards to make process analyses comparable. What are the "key ingredients" (KPIs) that you look for when analyzing processes? We already spoke about the levers of process in the context of steps 1 and 6 of the six-step approach, but have you defined and communicated what "good looks like"? What are the things that are most important for your organization when identifying and implementing improvements? Is it speed, capacity, or cost? Or any other criteria?

As you can see, implementing it can be quick or take some time. Either way, you always should have the bigger picture in mind that allows you to grow your program to other org units as quickly as possible. For this, I recommend documenting all the decisions that you made and also creating the artifacts that are mentioned above (such as the governance processes). This will allow you to communicate clearly and also review the foundations regularly and see if they are still a fit for your current and future situation.

11

Getting Started Summary

Getting started with process mining is a bit of a stretch, but then it is also nothing out of the ordinary when you have already implemented business process management as a discipline in your organization. It takes just the best of the analysis techniques and gives it a data-driven spin.

However, there are a few takeaways from the things around your process mining initiative that you should keep in mind before you dive into the example and the six steps in the next segment:

1. There is—as always—a lot of chatter around a new technology like process mining. Many people want to distinguish themselves from the rest of the herd (for example, on LinkedIn) or their competitors. But when you take a closer look at it, you will see that there are just nuances in the conversation.
   One big difference, though, is the technology between process and task mining. Both serve their purpose, and their use cases can be complementary. When using task mining, be prepared that you will add the aspect of trust to the equation, and you will affect your employees in ways that rub them the wrong way.
2. A different chatter is the conceptual differences between case-oriented and object-oriented process mining approaches. They differ in the way you construct your data model and how quickly you can apply changes

to your analysis (step 3 in the six-step approach) but do not impact the other five steps so much.

However, it adds a level of complexity to a project (just like automating data feeds from source systems) that might work contradictory to the objective of having numerous smaller projects with fast turnarounds to create momentum. Therefore, I recommend getting started with the simpler, traditional approach to process mining in the beginning.

Once you mature your process mining skills, setting up an OCLE-compliant data store makes a lot of sense and will add even more value to your analysis activities.

3. When selecting a process mining tool, keep in mind that these tools can do just so much (see the differences between task and process mining). To do a full process analysis, you will need additional tools that I will discuss in the next segment in detail: a data modeling tool, your process mining tool, a repository-driven process design tool, and a simulation tool.

   However, defining the requirements in detail for these additional tools will extend the scope of this book. And a process mining tool alone brings value in itself.
4. The enterprise software buying process is complex. There are many players involved, and all have their objectives, and there might be a lot of opportunity for miscommunication or even conflicting interests, so that you, as the person who is responsible for bringing in the new data-driven analysis capability, will have to have a clear vision of what you want to bring in at what time.
5. Speaking of implementation: Process mining is ideally an enterprise-wide tool because the benefit of seeing how your processes run in reality and identifying improvement opportunities small and large along the way is an immense benefit. I like to compare this to the situation that nobody in a regular setting would buy a car without a speedometer for driving on public roads. But then, process mining could easily be implemented for one group first and then rolled out to additional org units.

Either way, you should keep the enterprise in your mind and follow the four phases of strategy, process and organizational governance design, technology, and preparation of existing content.[27]

III

Process Mining in Real Life

12

Process Mining in Real Life Introduction

This is the part where the fun begins—you take control of the steering wheel and use the (process) speedometer yourself.

In contrast to other books, "real life" means that you can download example files and follow the six-step approach for successful process mining projects with your tools. As you will see, I am using the ARIS platform as an example, but it doesn't matter if you use a different set of tools ... you should come to the same conclusions. Please reach out to me if that is not the case (and where I made a mistake ;-)

However, process analysis is not confined to process mining alone, as you will see, and I am using two other features of the ARIS platform (design and simulation) that your tool might or might not have. For these chapters, take it as the start of a requirement collection to see what is available in mid-2025 in a mature tool stack.

What's in this section?

This section goes through all six steps of the process mining approach introduced above:

- An introduction to the fictitious medical device company that wants to do their first process mining project and see what they can learn from this analysis approach. This covers the first step of the process mining strategy.

- A look at the tools you might need for process analysis and to follow our example yourself.
- The creation of a data model as the foundation for all mining activities later.
- The development of a reference model for conformance analysis later, and an introduction to base analysis and simulation in case you don't have a process mining tool (yet) or want to supplement your mining tool with these features.
- The actual process analysis and the different techniques that are built into process mining tools.
- A short detour talking about dashboards as the main communication tool in process mining and what to think of when assembling them.
- Identification of process improvement opportunities and how you can prepare those ideas to an approval-ready status.

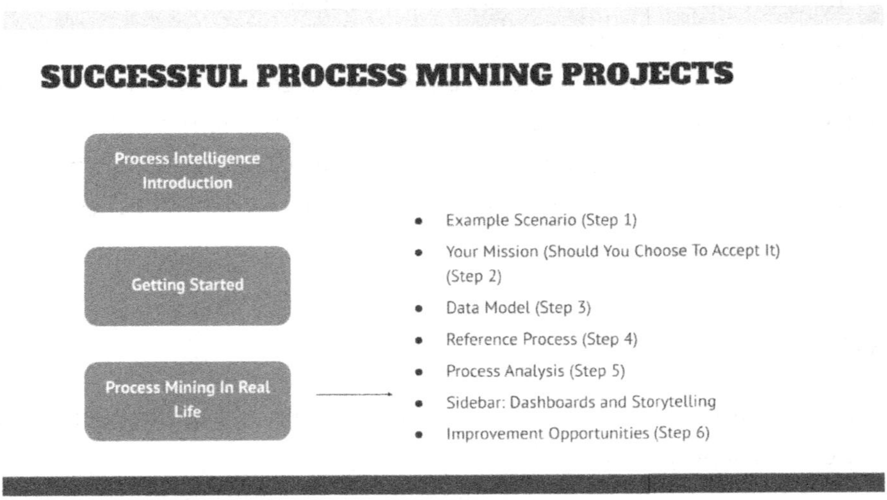

Fig. 12.1: Process mining in real life structure

13

Example Scenario

The example scenario for the rest of this book is a medium-size global medical device company named Heartbeat Medical Devices, Inc., with an annual revenue of 8 billion dollars and a global footprint.

The leadership of the organization has seen your latest presentation about process mining at a conference and loved the six-step approach and the promise of quick-turnaround, value-creating projects. Momentum is all for them, but they have concerns that process mining will not deliver the results that you promised. However, they are willing to invest a relatively small sum for a pilot that they want to do in one of their locations with the condition that it is time-boxed to a 6-week project because they would like to discuss and decide on if process mining is a technology that they should bring in when discussing next year's budget at the annual planning meeting two weeks after the project will end.

A Statement of Work (SOW) defines the scope and timelines of the project, and the SOW, as well as the project briefing (the slides in this chapter), are available for download on the accompanying website to this book.[28]

About the client

Heartbeat is a global company that is headquartered in Washington, DC, but has local headquarters in Frankfurt (EMEA) and Sydney (APAC). Additional "local" sales offices have been opened in Buenos Aires for South America and Kyoto to cover the strong customer base in Japan.

The company builds not only the medical devices themselves but also offers complete solutions that include software development for tracking the devices and their performance with patients that is also built in-house. This allows Heartbeat to establish a vertical integration and ecosystem up to automatic readings from the devices when a patient enters the office of an associated doctor. The software also connects to the patient's phones to give them alerts when the device (or the patient's health) operates outside the regular parameters so that the patients can be checked before something critical happens. For obvious reasons, there is strict regulatory supervision applied to the company, especially when it comes to data protection and the use of AI that affects patients.

The development of the devices and software happens in California, Croatia, and India, and the manufacturing happens in Canada, Mexico, and Chile.

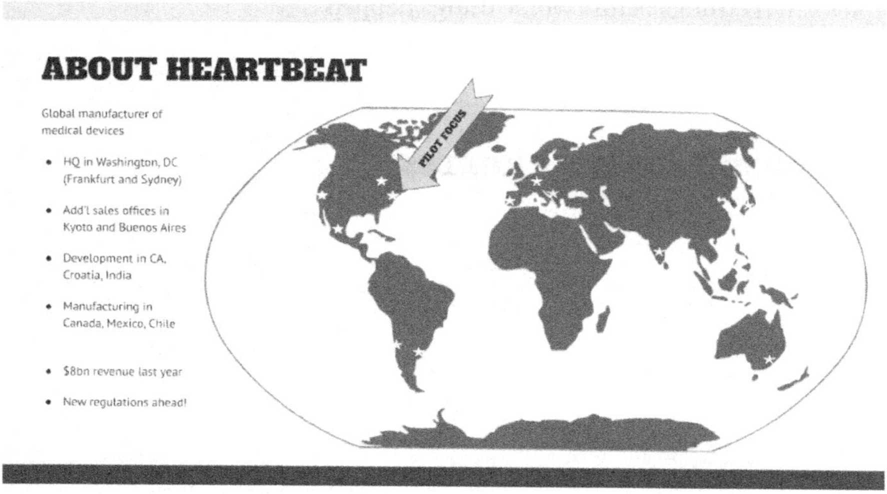

Fig. 13.1: Heartbeat Medical Devices global setup

The project

The pilot project will take place at the global headquarters in Washington, DC, and since this is a "test" project, it will cover the procurement process for this one location ("Procure-to-Pay" process).

In preparation for the project, the Business Process Management team has already mapped out the process in a diagram. The process has been reworked multiple times over the last couple of years, and the process owner is very proud of the achievement and development of the process—the team was able to reduce the number of involved systems to two (the procurement and finance systems), and they were capable of building integrations with vendor systems that then have access to Heartbeat's systems and will show up as "system" users, similar to the internal automation.

However, there is also the feeling that the process could be further improved, and some performance metrics of the process that are quarterly evaluated are not as expected. One promise of the process mining technology is to get early warnings when things might go South so that the executing roles can fix things before worse things happen.

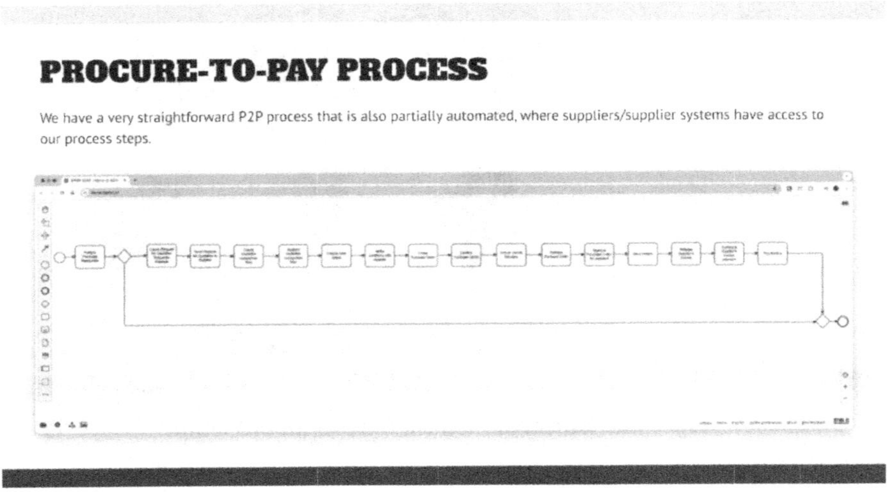

Fig. 13.2: End-to-end process for the project ("procure-to-pay")

In addition to that, the client team also has defined some project objectives as part of the presales work that led to the Statement of Work. The following figure shows the areas in an overview:

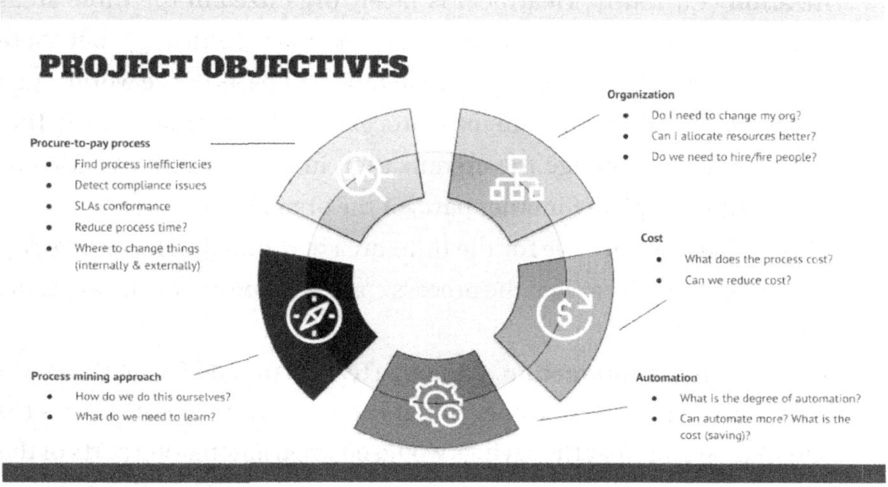

Fig. 13.3: Objectives and hypotheses of the project

- **Analysis of the procure-to-pay process.** This includes finding process inefficiencies, compliance issues, and conformance to service level agreements. Especially, three items are of interest:
  — The duration between "Create Purchase Requisition" and "Analyze Purchase Requisition" should not exceed 1 day.
  — Duration between "Analyze Request for Quotation" and "Create Purchase Order" should not exceed 7 days.
  — The financial manager role should just spend only about 1% of their time in this process—they are busy with other stuff.
  — The client team is also wondering if there are other industry KPIs they should look at.
  Furthermore, one of the key tenets of a process for Heartbeat is to increase the execution of any process in the company while keeping the quality of delivery up and the cost of the process low. They also want to know where to change things—internally and externally (when interacting with their vendors).
- The second area of the Project Mining analysis shall cover the **orga-**

nization. Currently, Heartbeat is nicely organized in the three areas above (corporate/sales, development, and manufacturing), but there are different levels of performance in different parts of the world when it comes to the procure-to-pay process, and the project shall find ideas on how to change the organization in DC that could be used as a blueprint/template for other parts of the organization later.
- This is also the objective for the third area of the analysis: **cost**. Today, the client has no idea what the process costs and how they can reduce the cost.
- The next area is **automation**. While the team is proud of what they have accomplished, they are wondering if there are more opportunities for automation and what the savings would be when automating parts of the process.
- Lastly, the project team is looking at the **approach for process mining** and wants to assess if they can do process mining projects themselves going forward and what the skills are that they need to learn.

Available data

Since this is a pilot project, no technical integrations shall be built, and Heartbeat's IT department has exported the data in CSV file format for download:

- an export from the procurement system,
- an export from the finance system,
- the salary of the individuals that play a role in the processes,
- a mapping of cases to product categories.

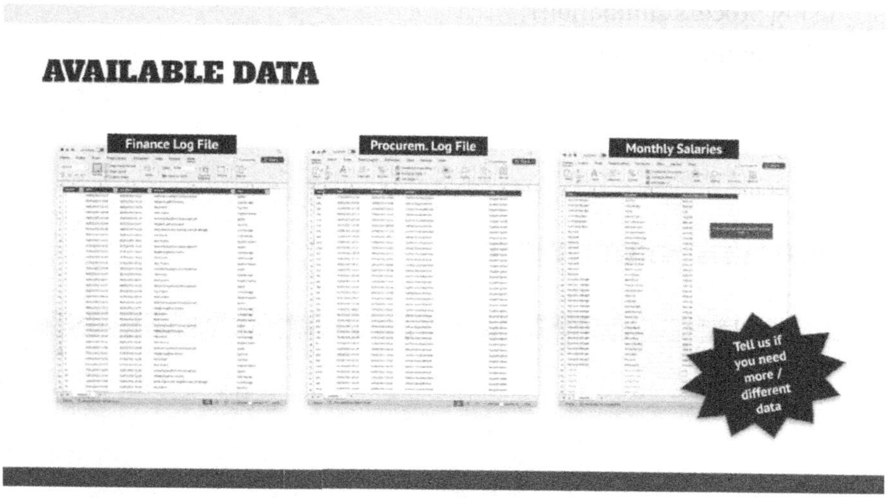

Fig. 13.4: Available data for the project

The project

As mentioned above, Heartbeat has a very tight timeline that cannot be moved so that they can meet their annual planning meeting date and decide if process mining is a technology that they want to introduce to the organization in the next fiscal year.

The expectation is that the project team will create a project plan that will be executed in the given time frame and produce the deliverables mentioned in the graphic below:

- Project workspace requisitioned
- Kick-off presentation, presentation notes (incl. objectives, etc.), high-level process model
- Transformed data, notes from validation workshop(s)
- Dashboards in process mining, analysis findings presentation
- Enhanced analysis findings presentation (include the proposed improvements and demonstrate the business case for at least one opportunity

using process simulation)

- Pilot findings presented

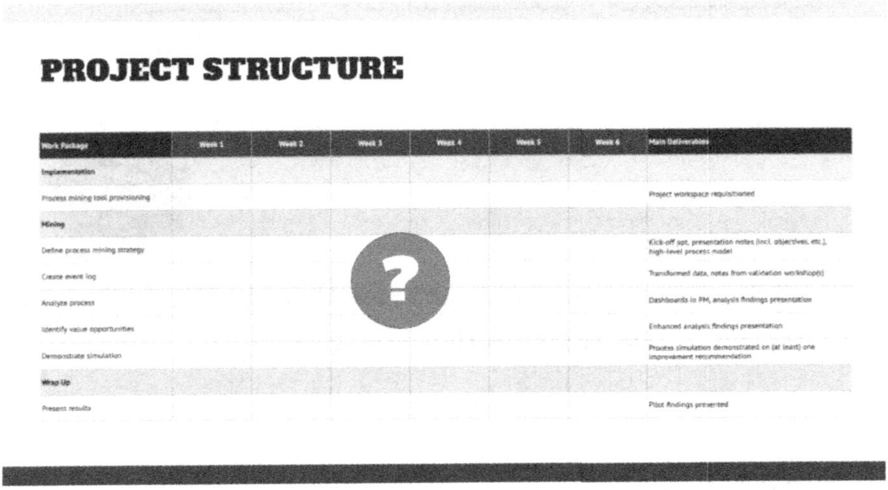

PROJECT STRUCTURE

Work Package	Week 1	Week 2	Week 3	Week 4	Week 5	Week 6	Main Deliverables
Implementation							
Process mining tool provisioning							Project workspace requisitioned
Mining							
Define process mining strategy							Kick-off ppt, presentation notes (incl. objectives, etc.), high-level process model
Create event log							Transformed data, notes from validation workshop(s)
Analyze process							Dashboards in PM, analysis findings presentation
Identify value opportunities							Enhanced analysis findings presentation
Demonstrate simulation							Process simulation demonstrated on (at least) one improvement recommendation
Wrap Up							
Present results							Pilot findings presented

Fig. 13.5: Project plan structure

In addition to this, the client has named three contact persons but is looking at the project team to identify additional resources needed to successfully accomplish the project objectives.

Heartbeat also expects regular status meetings and workshops during the project to be included in the plan, and the project team shall take over the coordination and facilitation of these meetings.

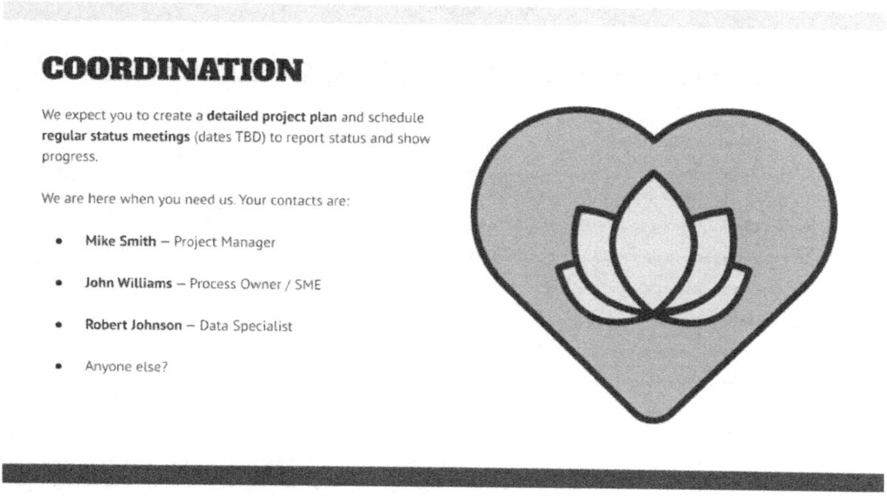

Fig. 13.6: Project coordination

Takeaways for the example project

This chapter covers steps one and two of the six-step process, and the main takeaways for the project are as follows:

- Heartbeat has very specific questions for the process analysis, which might require more analysis types than what a typical process mining tool includes. The project needs to give answers to all mentioned questions.
- The client is also asking for additional SLAs that could be added to the analysis and is looking for the project's guidance and industry expertise. Now, this can be a challenge if you are a tool or analysis expert, but I recommend that you take a look at reference frameworks, such as APQC,[29] that ship with KPI definitions. Alternatively (or additionally?), you also might want to ask the AI of your choice for some ideas, and you will be astonished by the quality of the KPIs and also the benchmarks for an organization that matches your client profile.

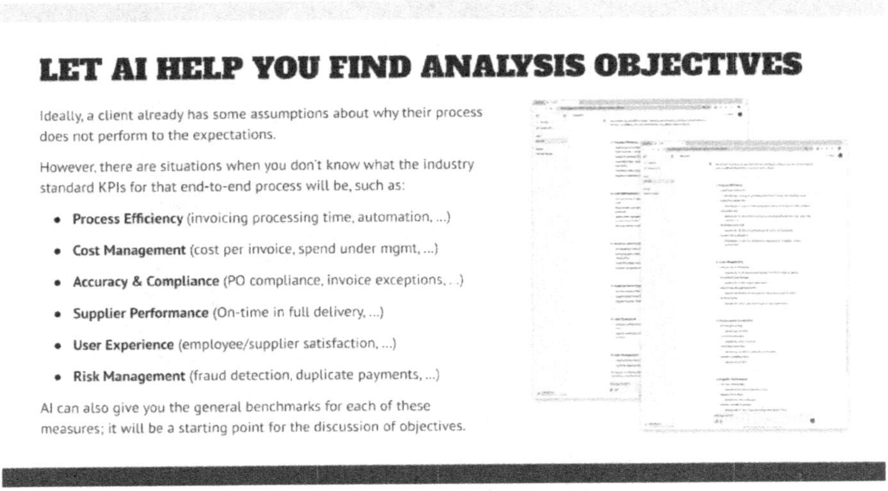

Fig. 13.7: AI to suggest KPIs and benchmarks

- Heartbeat has articulated the best possible execution of the process (capacity, time), keeping quality in delivery up, and cost down as criteria for a "good process." This might have an impact on the assessment of identified improvement opportunities.
- The process is modeled in BPMN (good), but is very simplistic. There will be more information available when the initial import in the process mining tool is done, and the diagram will have to be enhanced in the modeling tool. However, keep in mind that they think that they have already optimized the process as much as possible—this can become a "personal conflict" if they believe that you are "smarter than they are," and potentially the client will resist your approach, findings, and recommendations ... Tread lightly when communicating.
- After a closer look at the deliverables in the SOW and the process structure template, it shows that the project needs to deliver at least one simulation, even though it was not mentioned explicitly in the project objectives listing.
- The timeline is "challenging." Therefore, the number of iterations

should be limited (maybe one or two maximum).

- This project will not build technical integrations to the source systems but will use exports from them. This might result in missing data that might be needed for an analysis. If that happens, the question that cannot be answered will be captured as a "parking lot item" for the next iteration of the analysis of this process in the future. Clearly communicate this to the client.
- Communication with the client is key. They have named three contact people, but those do not cover the major roles in the six-step approach. Which ones are missing?
  In addition to this, the short timeline requires a clear approach to what happens when—above and beyond the regular status meetings. You need to clearly lead the client through the six-step process and over-communicate. As in any other project, the management of people is almost more important than objective outcomes. Remember: one of your objectives is that Heartbeat will decide to add process mining to their analysis tool stack and budget for it in the next FY plan.

14

Your Mission (Should You Choose To Accept It)

As you might have already read, the character of this book is "hands-on," and I would love if you took the files provided and tried to recreate the results for yourself. And maybe you find something else that I have not mentioned here... Please let me know :-)

But what tools do you need?

The graphic below shows Heartbeat's questions from the previous chapter in an overview and maps them to different analysis tools:

- **Mining tool**: the tool why you might have picked up this book and that you are evaluating for your organization.
- **Data tool**—required for transforming the source data from the execution systems into tables that can be imported into the mining tool. Most process mining tools come with some capabilities for data transformation, but those might be limited, and in almost all cases it requires some form of coding, which is not in the typical skill set of a business analyst.
- **Modeling tool**: One big step towards improving the process in scope is to make it visible and model it in BPMN. This allows the upload of the

reference process into the mining tool (for conformance check purposes), but also the download of discovered processes to get you started with an "80% process" if you do not have a diagram available. You can also run base analysis in these tools by running reports and other features.

- **Simulation tool**: You can use simulation as part of your process analysis ("Does it really play out that way when I look at that other variable?") and to validate your improvement opportunities ("Will the proposed change yield the benefits that I have calculated?").
- **Collaboration/publishing/office tools**—the usual suspects that you might already have in your organization: an office suite (Google, Microsoft, LibreOffice, etc.). A collaboration tool to keep you on track and a publishing tool.

  For the latter two, I recommend Notion or its open-source counterpart, Affine. We are using both for our podcast, and for the rest of this book, I will use Notion. It has a free personal plan, and more importantly, it can share pages as a web page.[30]

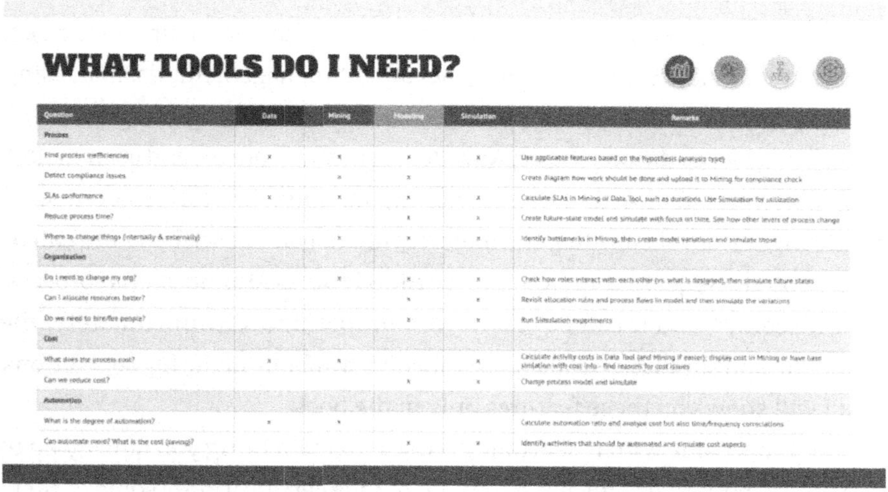

WHAT TOOLS DO I NEED?

Question	Data	Mining	Modeling	Simulation	Remarks
Process					
Find process inefficiencies	x	x	x	x	Use applicable features based on the hypothesis (analysis type)
Detect compliance issues		x	x		Create diagram how work should be done and upload it to Mining for compliance check
SLAs conformance	x	x	x	x	Calculate SLAs in Mining or Data Tool, such as durations. Use Simulation for utilization
Reduce process time?			x	x	Create future-state model and simulate with focus on time. See how other levers of process change
Where to change things (internally & externally)		x	x	x	Identify bottlenecks in Mining, then create model variations and simulate those
Organization					
Do I need to change my org?		x	x	x	Check how roles interact with each other (vs. what is designed), then simulate future states
Can I allocate resources better?			x	x	Revisit allocation rules and process flows in model and then simulate the variations
Do we need to hire/fire people?			x	x	Run Simulation experiments
Cost					
What does the process cost?	x	x		x	Calculate activity costs in Data Tool (and Mining if easier); display cost in Mining or have base simulation with cost info - find reasons for cost issues
Can we reduce cost?			x	x	Change process model and simulate
Automation					
What is the degree of automation?	x	x			Calculate automation ratio and analyze cost but also time/frequency correlations
Can automate more? What is the cost (saving)?			x	x	Identify activities that should be automated and simulate cost aspects

Fig. 14.1: Tools needed for process (mining) analysis[31]

Analysis tools

Obviously, there are virtually unlimited combinations of potential tools, but in this book I am using the ARIS tool suite for mining, modeling, and simulation and KNIME for data modeling.[32]

KNIME is an open-source tool that can be downloaded from knime.com.[33] It allows you to model transformation in a graphical environment by creating workflows in which each node makes one change. The tool shows you if you have configured the node correctly by showing a small traffic light icon, which will help you find the node in the workflow (which can grow so huge that I christened one client data model "the beast," and one of my consultants refused to make changes because he was afraid to screw it up) and to debug it.

It also "self-documents" because it allows you to add comments to each step in the workflow, as well as annotate parts of the process and highlight the different areas in scope. In addition to this, it also allows you to create reusable components, similar to the subprocess functionality in BPMN. This makes a workflow more readable for the casual user while still being able to drill down into the details in a new tab.[34]

ARIS is a proprietary tool for business process management and is a very mature tool that provides modeling, document management, dashboarding, simulation, automation, and process as well as task mining features.

It comes in three editions (Basic, Advanced, and Enterprise), and even though the Basic version of ARIS Process Mining has some limitations compared to the higher-level editions, the majority of analyses described here can be done. You can request your 90-day free ARIS Process Mining at aris-process-mining.com,[35] and it can be extended multiple times. Unfortunately, the modeling and simulation capabilities are only available in the higher editions, and I will show you these features only in the book.

Both tools come with e-learning courses that will bring you quickly up to speed with both tools and also allow you to take certifications (some of them free of charge), which you can add to your professional profile.[36]

But, of course, these are just my choices of tools, and with a bit of research, you will find many open-source or proprietary tools that might fit your wallet. I encourage you to still take the example files and recreate the analysis that I do with the tools mentioned above.

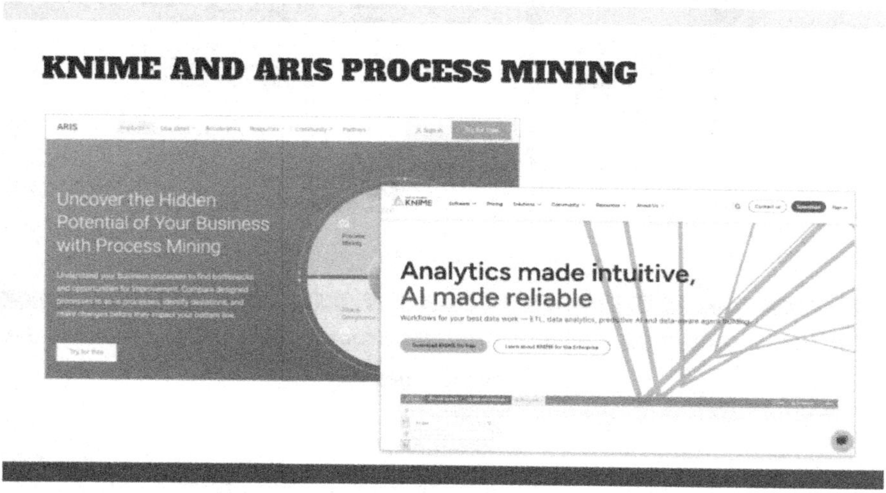

Fig. 14.2: Two of the tools used in this book

Where do I get the necessary files?

As mentioned earlier, there is a companion web page at whatsyourbaseline.com/heartbeat where you can download all the necessary files that are used here, as well as the "solutions" for different steps in the six-step process.

If you find something missing, please reach out, and I am happy to take a look.

PROCESS MINING EXAMPLE FILES

whatsyourbaseline.com/heartbeat

Scan the QR code or go to whatsyourbaseline.com/heartbeat[37] *to download the needed files*

15

Data Modeling

Data modeling is the step that might take the most of your project time. Initially, you might think that this is a simple, one-time exercise, but in my experience it is something that you will do almost until the end of the project.

Why data modeling for process mining?

Well, the obvious answer is that data is the foundation for process mining as a data-driven analysis method. But there is more besides making data available for your tool of choice:

- You might want to **normalize** all your data by removing redundant records in a table or removing columns from other tables that hold the same information. The idea is that you have one piece of information in one place and then relate the information to each other.
- You also might want to **standardize** the data, which means that you can compare apples to apples later. This can be helpful when you calculate KPIs later as part of your analysis, and, for example, being able to calculate activity cost. You might want to have a common scale, such as "cost per minute."
- You also might need to **change the data type**. For example, when you want to calculate the duration of event data or extract the weekday or

month from a timestamp (to show a graph of busiest weekdays in this process), you need to transform whatever the input format is into a time format.

- The main point of data modeling, though, is to **transform the source data**, which can come from different systems in different file formats, into a format that your mining tool understands and can work with.
- In addition to this, you also might have to do **analysis outside your mining tool** (because it might simply not have the functionality that you need, such as a 2-factor analysis), and these calculations and analyses should always come from the same source data set.
- And lastly, you might want to make your **data more understandable** by renaming columns into human-friendly names or sorting an output table in a "logical" manner so that it is easier to understand.

Either way, you need to be able to justify everything that you do with the source data. You want to make the transformations visible, and you can walk your stakeholders through what you did to get their buy-in.

Data modeling for process mining is structured into four distinct activities:

1. Data cleansing of the source data
2. Creation of the initial activity log (and upload into the process mining tool to see if you can get an initial discovered process and the base statistics)
3. Calculation of metrics needed for your analysis
4. Export the event log

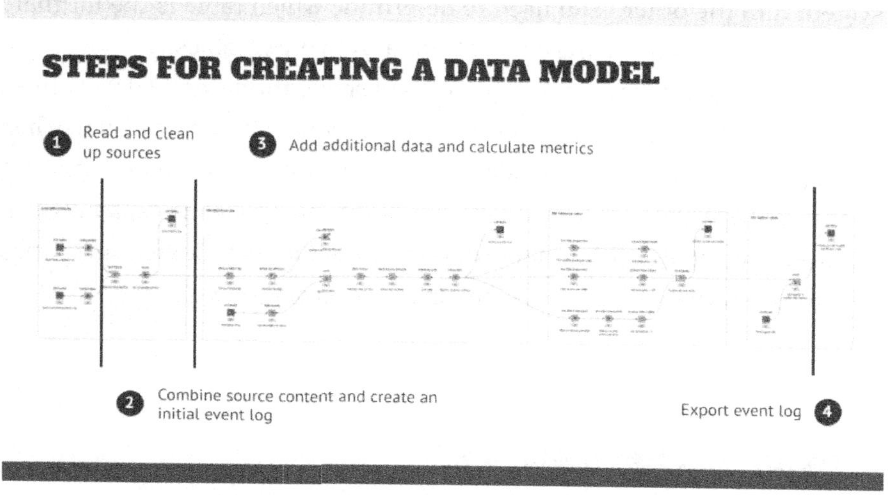

Fig. 15.1: Steps for creating a data model

As mentioned above, you can do this with any tool that you like, but I will use KNIME going forward because it demonstrates the thinking of what needs to be done visually. I highly recommend visual data modeling over programming. Some tools have even created KNIME nodes to connect it directly to their process mining tool—the way to go forward, IMHO.

Data cleansing

The first step you need to do after you have determined which system data you need for each step of your process in scope is to clean the data. This is done in multiple areas. that you will have to cover.

Understand the process and data structure

I mentioned it multiple times above, but you *really* must understand the data for your process mining process, and unfortunately, you might find redundant information in your source files—for example, if you don't use a case-oriented tool like Salesforce or if there is data replication from one

system into the other. You need to determine which table is the ultimate source and later remove the duplicate data from other tables.
It can be very helpful to create a diagram that shows the different data entities and their attributes, as well as the applications where they come from, when you explain this to your stakeholders later. If you use a mining tool that is part of a larger BPM platform, you might even have data modeling capabilities in that tool. Figure 15.2 below shows example data models depicting entities and then the attributes of that entity in a second diagram.

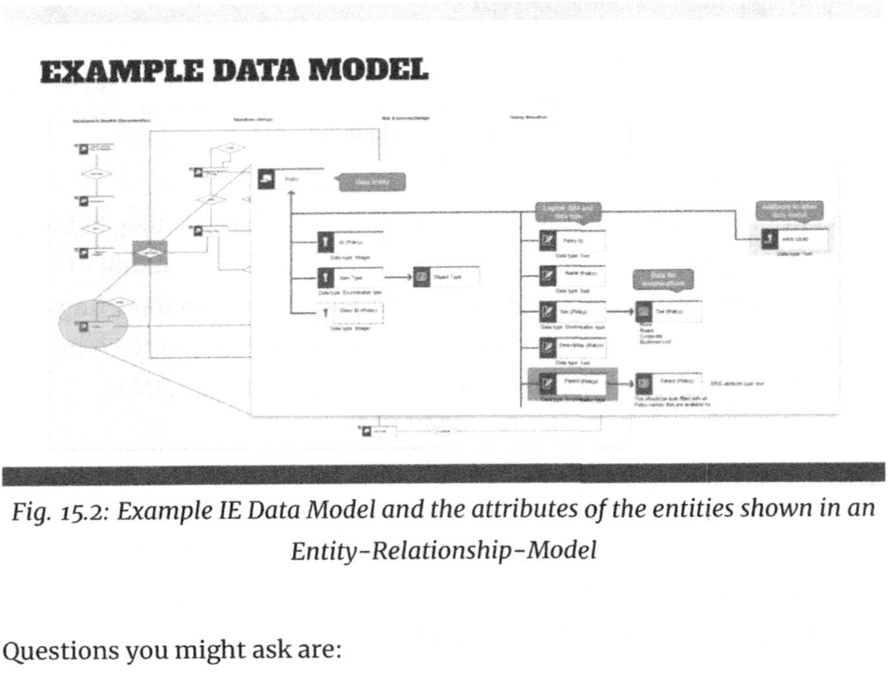

Fig. 15.2: Example IE Data Model and the attributes of the entities shown in an Entity-Relationship-Model

Questions you might ask are:

- What are the **activity logs, and what are dictionaries/additional information** needed, for example, the "translation" of abbreviations into "full words," which will make things easier to understand for the analyst or stakeholders?
- What is the **input structure**—which table is the main activity table and

which ones are the auxiliary tables with additional information that should be added to the overall data model later (and which information in those are relevant)?

- Is it advisable to **split** any of the tables into multiple, and where do I need to **join** tables? You can accomplish this by using the Column Filter node in KNIME multiple times after you've read the source file and have two branches that only include the columns that you require for your identified purpose.
- Do you need to ask your IT friends for a different, more **optimized export from source systems**, which will make your data modeling easier and provide fewer opportunities for errors?

File and other formats

In this step, you look at the format that your data is provided in.

- Is it in **flat files** like a CSV or Excel file, or do you have to connect to a **database,** and are you sure which table to connect to?
- And when you get the source export, check if the **encoding** is known (such as UTF-8 when reading a CSV file) so that no data will be omitted when you read the content or comes in "scrambled" with unknown characters.
- You also must determine the structure of how you **file your source files and where to store your exported files**. The golden rule is that you will never touch or overwrite the source files but rather export changed data in separate files.
- The next topic is where to store these files—locally **or on a shared drive**, such as SharePoint? Can everyone access the data and have the correct permissions? Does your process mining tool need to access the files from somewhere, or will you upload changed files manually every time?
  If you go the shared route in a multi-user environment and want to read/write them from local data modeling tool installations, then you need to set the read/write locations as relative locations in the tool and not as "hard locations." How will this be different if you use a shared

installation of your data modeling tool?

- Are your **columns in the right data format**? When you import a CSV, for example, then all columns will be pure text ("string"), and you will have to determine for each column what the "real format" needs to be: "local time," "integer," "double," and so on.[38] You can use different nodes to change the data types in KNIME, such as the String to Date & Time node. Please keep in mind that many process mining tools will require CSV files for manual upload, and you will have to do the data type mapping in the mining tool again. It still makes sense to work with the correct data type in your data modeling tool so that you will be able to calculate things later that will be imported into your process mining tool as separate columns.
- You also want to make sure that the **columns in the different tables have a unique name** so that they don't show up as "duplicates" in your mining tool when it comes to configuring widgets on a dashboard, and you will be unsure which one of the multiples is the one that you need ... your data model mentioned above can be a good map for future reference.

Human friendliness

One of the things that bothers me all the time is that columns in a data file are named in a non-human-friendly way. This can be the column header or the values themselves. To fix this, I recommend the following steps:

- **Rename columns** ("LOC" becomes "location," for example) using the Column Renamer node.
- Look into the values in the columns and do **string replacement** activities and remove/change special characters and umlauts where needed (String Replacement node).
- Apply a **consistent case sensitivity** and standardize on a consistent uppercase/lowercase for columns and values.
- Create **dictionaries** that you will use as a lookup source for the proper, human-friendly name (for example, to replace product codes with the standardized product names). You can do this by creating a separate

spreadsheet or using the Table Creator node in KNIME and then joining the two tables at some point and adding another column.

- **Remove columns that you don't need** anymore (Column Filter) and sort the sequence of the columns so that they make sense for a human who looks at your data (Column Resorter node).

Missing data

Missing data can come in different forms—truly missing data/empty cells, unwanted characters, or truncated data.

- Unfortunately, you might run into the situation that your data set might have **incomplete data**, and you need to decide what to do with it. Will you have to replace the missing data with a static value for text and numbers? Do you need to filter out the rows that have missing values?
- Your data also might have **unwanted characters** like leading or trailing whitespace, tabs, line breaks, or duplicate whitespace in the values. This might prevent you from transforming columns into another data type, and you want to clean up these artifacts. Luckily, there are nodes in KNIME that will do these things for you, but please make sure to check if they have done it correctly—a good practice is to not check every box in the String Cleaner node and to hope for the best. Rather, do each cleanup as separate nodes and check which character cleanup is left to do, and use a new String Cleaner node for the next cleanup. Being slow and systematic here will pay out in the future.
- You might see **truncated data**, which could be intentional (the data is stored with a limited number of characters) or be an unwanted artifact of the export settings from the source tool. Check with your IT folks if the export is correct, and if the truncated data is correctly exported, you might think if you want to create a new dictionary and do a lookup to give the data a human-friendly value.
- And lastly, it **might be OK to have missing data**. For example, if your process mining activities have start and end timestamps for an activity,

a missing end timestamp can indicate that this is a case that is currently in progress and the data was exported while that activity has not been completed.
However, you might want to check if that is really true or if there are later steps with a newer timestamp for this case, and you are missing the end timestamp. In that case, you need to decide if you fill the end with the start timestamp + x minutes so that your calculations that expect an end timestamp don't fail.

Filtering

Depending on the export that you get from the source systems, you might not only get tables that you don't need, but you also might have columns and values in the rows that you don't want. Applying filters will help you to clean up these things quickly.

- We already spoke about **removing columns** from a data set above and giving them unique names, but another application could be to **filter unwanted steps** from the data set. For this, you can use the Row Filter node and define the criteria for it in its configuration, such as a string or using wildcards.
- There is a similar node available called Duplicate Row Filter, and it does just that—it **removes rows that are duplicate** to others in one or more criteria (like time or a string value). You can decide which ones to keep (the first or last value, for example). This can be useful when you are calculating duration in some situations or when you need a list of unique values ... for example, when you want to create a dictionary for reference lookup later.
- You should also **filter out nominal data** if your export includes static text or category names that are not part of the data set but are added by the report from the source system as a courtesy for the human reader.
- A last use case is the **filtering of statistical outliers**. For this, you obviously need to have a definition of the data range that is expected and

acceptable, and it could cover timestamps (everything before a certain date or where the duration is greater than x) or values (orders of less than x items or where a location shall be excluded because it is out of scope of the process mining analysis).
Please note that you will also do filtering in the analysis later in step 5 of our approach; the difference is that you don't want to remove data before it is needed now. Or, to say it in other words, filtering in the process mining tool is to highlight a certain point that you want to make to verify/reject your hypothesis.

Times

Working with time is tricky because we typically have 8-hour workdays (and not 24-hour workdays), holidays, or other times that should not be used for calculations—for example, weekends.

- The foundation for all of this is a **consistent date format**: do you use the European date and time format or the American (dd-MM-yyyy vs. MM-dd-yyyy; AM/PM vs. 24-hour time format)? What are the dividers (hyphens, periods, slashes)? And so on.
  You want a consistent format for all dates in your data set, and you might want to check with your process mining tool which formats it accepts—most tools adhere to the ISO 8601 standard, which also defines what characters in upper- or lowercase mean (there is a difference between "M" and "m," for example).[39]
  This can be tedious and might require multiple nodes and actions to transform time into strings, manipulate them, and then transform them into time again.
- Ideally, you want to have **separate timestamps** for the start and end of activities, but there are ways to approximate end timestamps from your data set (a complex approach).
- When working with time, you might find **negative time** (either when the end comes before the start timestamp or when you are calculating time).

The first is obviously an error in the data set, and you have to decide what to do with that row (change, filter out), while the latter might require checking your calculation logic.

- And lastly, sometimes you find those sneaky **non-UTF characters in timestamps** as well, which will throw you off :-(

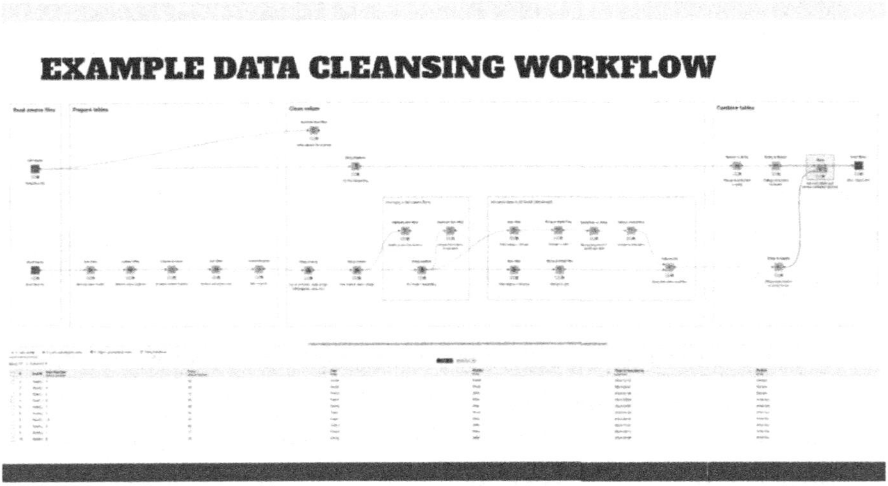

Fig. 15.3: Example of a data cleansing workflow[40]

These are just a few examples of things you have to do when cleaning your data. For a more comprehensive introduction to this topic, I highly recommend Benjamin Greve's book "A Beginner's Guide to Clean Data"—he covers the things above and digs deep into the necessary preparation for other use cases, such as machine learning or text analysis.[41]

Create an initial activity log

Once you have understood your data structure and cleaned up the tables, it is time to assemble a first activity log that you upload into your process mining tool.

To make process mining work, it does not take a lot of different data—you just need a few different things, as we've seen above: a case ID, activity names, and at least one timestamp for each instance. To accomplish this, there are a few activities that you might have to do.

Merging data sources

If you get your activities from different data sources (like in our Heartbeat example), you have to bring them together. There are two different ways that you will have to keep in mind:

- Do your data tables have the same structure, and you just need to add more rows from one table to another? Then you will **concatenate** the tables using the node with the same name.
  Maybe you have to create additional, empty columns to make this happen in both tables, and KNIME has the Constant Value Column node to accomplish this.
- If you need to map values from one table to another by using a common field (as in our data cleansing example when we added the price to the other table using the item number as the shared field), then you can use the **Joiner** node. This node allows you to set different methods of joining (only matching rows, only matching to one of the tables, and also the option to merge joint columns). This is the join command in SQL, and KNIME will create combinations depending on your settings. This can lead to the creation of new rows, so you want to take a look at how many rows you started to join with and if the aggregated number of rows is the same after the join.
  For this operation, you do not have to have the same columns; the tool will add all columns by default (and add a "right" to the second table columns). Conveniently, you can remove the not-needed columns in the configuration dialog of the node so that you don't need to apply a separate Column Filter node after you join both tables.

However, I like to add a Column Filter anyway temporarily and remove all columns from the combined table that are not case IDs, activities, or timestamps, and then save that table as CSV using the CSV Writer node. That is the file that you then can upload into your process mining tool to see if the base activity log works.

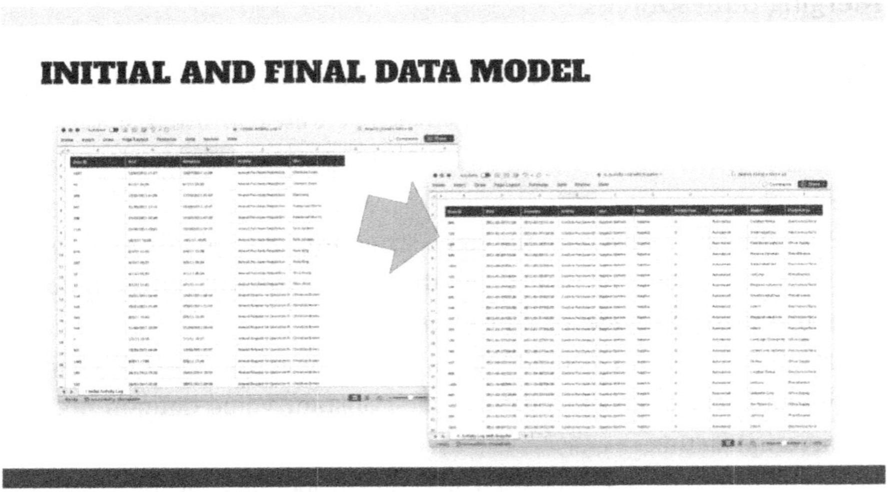

Fig. 15.4: Prepared data model ready to be uploaded into Process Mining tool

Activities

But what happens if you need to add more transformations to your minimal data set so that it shows up as you like? I have seen a couple of scenarios that you might encounter as well:

- There is **no case ID** in your data set. This is a serious issue because without this, process mining will not work.
  But do not fret; there are ways of creating a case ID. The easiest way might be to see if your tool does this automatically (they typically distinguish

between a case and an activity perspective, and you would choose the latter). The idea is that you will have to create the same columns in the different tables. If there is a reference to a different table as mentioned above,[42] you know that it is the same case. The tool will assign case IDs automatically.

- Sometimes the **activity names** are *not really* a 'speaking' activity but might be stored in the source system with added information, such as the name of the process or phase within a larger process. If you encounter this, you can use the Cell Splitter node and define a character, such as a hyphen (or number of characters, such as "after the 5th character"), where KNIME shall split the value of the cells into multiple columns. Thereafter, you can delete the unneeded column and proceed.
- Lastly, you can also use the String Manipulation or String Replacer nodes and have certain words replaced with others. You might need to do some filtering, work in multiple branches, and bring the branches back together after you've transformed the names so that you have a consistent naming, and "create invoice" and "invoice creation" will have the same name, for example. You will need this for harmonizing names for the conformance checks in your process mining tool if your tool does not support different names for activities in the reference process model and the data set. In that case, you would need to harmonize the names as part of your data modeling so that the conformance check understands that step A in the BPMN diagram is the same as step A of the data set.

Upload of initial activity log into Process Mining tool

Once you have your initial activity log created and exported as a CSV file, you can upload it into your process mining tool of choice.

All of them work similarly, but since I am using ARIS Process Mining here in this book, please find a screenshot of the data section for illustration below.

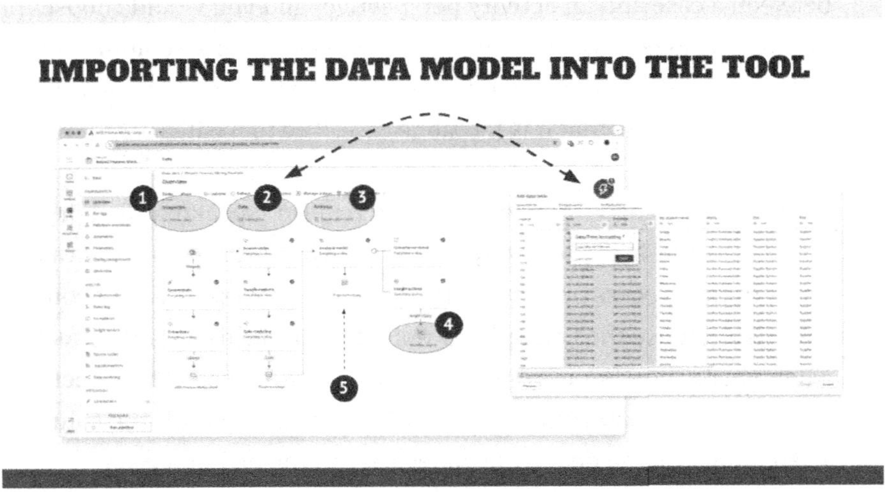

Fig 15.5: Column data type configuration in ARIS Process Mining

One thing I like in ARIS Process Mining is that it shows all data-relevant areas in one graphical overview, and these are reflected as menu items in the left-hand column (shown in Figure 15.5)—a clear user interface that also shows the status of the steps in the graphic so that you can see if a step failed.

1. **Integration**: ARIS Process Mining allows you to connect to source systems directly, and this section shows you the connections and if the extractions from the connections worked.
2. **Data**: In this section you choose the source tables (and this is where you upload your CSV files), do transformations in SQL-ish code (not needed since we use KNIME), and do data modeling—a bit of a misnomer: here you combine different source tables and indicate which of the uploaded tables is the activity log.
3. **Analysis**: Here you can do multiple things—calculate additional fields in ARIS Process Mining as part of your analysis model (see below for when you should do this). You also need to trigger recalculation of a data model if you made some changes.
4. **Analysis (continued)**: Here you can configure the compliance check

(the comparison of a reference model in BPMN format with the data set) and triggers for a workflow engine, such as thresholds for the duration of activities or a process instance. This allows you to interact with the running processes in the runtime systems, as mentioned in the "Process Mining in the Solution/Process Lifecycle" chapter.

5. **Analysis (continued)**—this marker indicates that you need the Analysis Model box executed (green check mark), and you can go to the analysis projects.

When uploading a CSV file in step 2, all numbers, dates, times, etc. will be reset to the "string" data type. You need to map the columns of your file(s) to the right data type. If there are inconsistencies or the mapping is not correct, ARIS Process Mining will color-code the columns and cells in red so that you can see where you need to change things. You also might want to check if there are alerts that some records have been filtered out during the upload, which, for obvious reasons, need to be debugged as well.

Then you need to load the data as a separate step so that ARIS Process Mining processes the uploaded and mapped files. Depending on the number of records, this might take a while.

After the tool has processed the data files, go to your analysis project and create a "Process explorer" analysis (a standard one-click analysis) and see if a process diagram with all cases has been created. In our Heartbeat example, you should have 608 cases available, as shown in the upper right-hand corner of the tool (see figure 15.6).

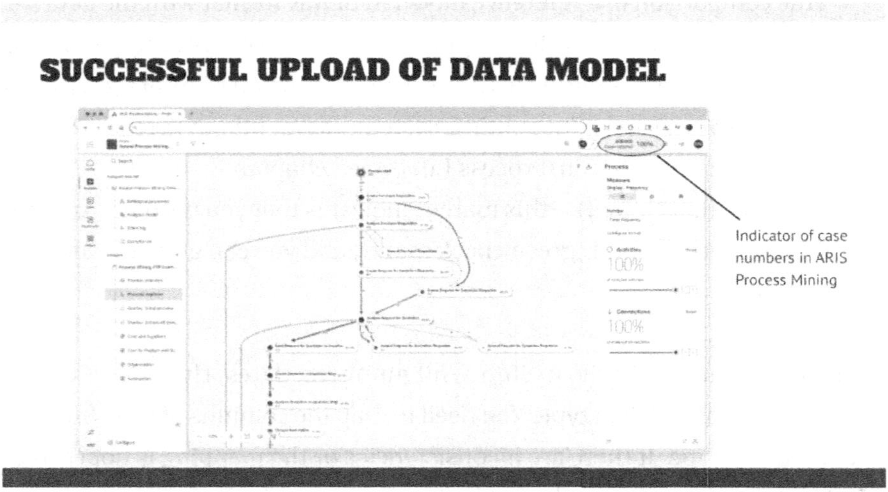

Fig. 15.6: Successful upload of data into process mining tool

Calculate additional metrics for your analysis

Congratulations — you have created your first activity log, and the tool works. This alone will give you some ideas about the process, but you will have to do more data work, and most likely this will lead to multiple iterations where you will upload your data over and over again.

Define the needs for additional analysis

The main triggers for additional calculations are the hypotheses that you've discovered in Step 1 of our six-step approach. For Heartbeat, this is nicely listed in the "circle slide," where they articulated what they want to know from the analysis. The client also gave hard SLAs that need to be checked, such as utilization of roles or duration between two steps that are not directly following each other.

At this time, you also might want to look at your process SMEs (and AI as well) to see how typical scenarios are processed and determine which additional calculations you might have to do. For example, in an incident management process, the client might have told you that they want to know how many incidents get escalated from a Level 1 to a Level 2 Support, but how many of those get further escalated, and which ones are solved by the Level 2 Support? How many of those get re-escalated back to L2 because the "solution" did not solve the problem?

2-factor analysis

Another example might be if there is a relationship between two factors. In one project with a bank, we looked at the due diligence in their projects, and one would assume that you put more effort into deciding if you should give out a significantly higher loan amount than your "standard loan." For this, we did a 2-factor analysis and looked at the loan amount and the cost that was accumulated for the due diligence activities—these were not only process costs but also included costs for services that were bought by other vendors (in these cases, studies to determine building feasibility).

To analyze these relationships, we created plots in KNIME that showed the cases using the Scatter Plot nodes and looked for anomalies. One would expect a 45-degree slope emerging (the higher the loan, the more effort), but we saw horizontal lines, which was interesting and rather showed "categories" of due diligence independent of the loan amount. However, there were many calculations (using filtering, joining, and Math nodes) to get the cost put together by case.

Long story short, you might need to do more analysis, which will only take place in the data modeling tool and is only tangentially process-related but provides the necessary context to analyze what you see in the process data.

Calculate KPIs

There are two more cases I'd like to point you to:

- I mentioned **cost calculations** already. However, this can be more than simple aggregation. Occasionally, you need to determine cost by category, for example. For this, you can use the Pivot or Group By nodes, and if you need to further transform a pivoted table, KNIME provides an Unpivot node.

- A trickier aspect might be the **calculation of business days** (as is the case with any time calculation)—if you have an SLA for a step, it is typically meant in business days and not calendar days, the standard that process mining tools use to calculate duration.
  If you are lucky, your tool allows you to create a calendar schedule that will do this for you, but for others, you will have to work with the Date & Time Part Extractor node and extract the calendar week of the two steps that you are interested in and extract the calendar weeks as a number from the dates. Then you subtract them from each other, and you will see if there is more than 0 that you have one or more weekends between them. You then need to calculate the duration from the two timestamps in days and subtract the weekend days (or additional holidays) from it.
  Furthermore, keep in mind that if you have something done on a Thursday afternoon and the activity was finished on Friday morning (less than 24 hours), it will show up as 0 days duration in KNIME. #sigh

Calculations in the process mining tool

For obvious reasons, you can go back to your data model and do calculations in the tool. However, it might be worth looking into the features of your process mining tool.

While I was saying that most tools require you to do SQL or other programming, some, like ARIS Process Mining, have additional wizards built into their tool. This will help you to calculate new fields that you can use in a dashboard widget later. You can see three types of calculations:

- Creation of new fields that aggregate data more easily than doing this in KNIME. One example is the calculation of the case cost ("pull-up" from the individual activity costs), which is a simple dialog in ARIS Process

Mining (shown below). This will save you some headache.

You can also find pre-built capabilities to create case IDs, etc., in the tool.

FIELD CALCULATION IN MINING TOOL

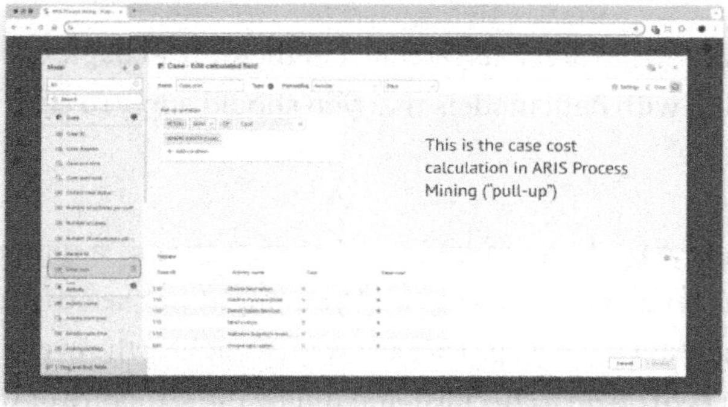

Fig. 15.7: Case cost calculation in ARIS Process Mining

- The second type of calculation is a bit more complex and uses the wizard interface to calculate a specific question. This is done in an if-then-else interface that allows you to click things together. However, it does not help in figuring out the correct logic for your question, and if you are a bit "logically challenged" (like me), this can be a frustrating exercise. It helps if you note the elements you need for the calculation on a sheet of paper before you dive deep into the wizard.
- Lastly, AI is built into the tools, and you can use it to have the programming snippet created that you can paste into the calculation wizard as needed.

In either case, this will force you to remember what you calculated where and have this documented as part of your project documentation. If you don't do this, someone might change something in a future version of the initial data model in KNIME and even though the resulting event logs load perfectly into the process mining tool, your in-tool calculations might break. You will be glad to have a list of the in-tool calculations for debugging somewhere.[43]

Data modeling best practices

I alluded to this in the previous section, but there are a couple of best practices when working with data models that you should apply to each of your data models.

- **Structure your data model** in different sections: a data model ("the beast") can be intimidating, and you might not see the forest for the trees and want to give up. The best way to not lose the overview is to note the things to do on a checklist and follow the structure of this chapter. Only when you've solved the problem at hand, move on to the next one. This will make sure that you don't forget anything.
- **Create slides that explain** how you got to the calculation. As mentioned above, in one project I had a data set that did not include start and end dates of each activity. However, you could calculate different approaches and then show the options to the client during the data validation. A single slide that explained the concept and options behind that was invaluable.
- **Count rows** during joins and other actions where you bring multiple tables together or filter. Depending on the settings of the nodes, you might create additional rows. Luckily, KNIME shows you the number of rows and columns in the preview, and I like to add a workflow annotation with that number at the outgoing connection to the next step to always see them in the workflow. This is useful when you filter, do transformations in each branch, and then join them together again. Obviously, you should not have more rows than you had before (fewer might be OK if you need to do some row filtering, though).

- **Clean up your model**. I mentioned it already, but group pieces of the workflow together and add a workflow annotation (box) around them to indicate that these steps belong together, as shown in figure 15.1 on top of this chapter. You could also create components and reusable meta nodes in KNIME, but that removes the overview of what happens in the workflow.
  However, if you are rolling out KNIME as your standard data modeling tool in your organization and you want collaboration with others, then this might be a good idea. On top of this, you should establish a version management system for the shared workflows.[44]

Takeaways for the example project

Data modeling happens in iterations, and for our project (as you can see in the solution workflow), it happens in four phases:

- Data cleansing and creation of the initial activity log
- Calculation of employee cost and joining into the main data set
- Calculation of the degree of automation (as an additional analysis step later)
- Joining of the product area into the data set to find cost-related improvement opportunities

Keep in mind that your hypotheses drive the effort you need to do (besides the "standard" analysis KPIs that your process mining tool calculates automatically), and you will have to have an answer for each of these hypotheses.

What you can control, though, is the number of iterations on the data model. This allows you to dedicate additional calculations as a "next project iteration" item if you are running low on time. Remember, you want to have short-turnaround projects to create momentum and not long-running projects that end in an underwhelming result.

Another item that the client in our example project went through is the walk-through of the data model so that they could understand how the data was transformed, and they gave their approval to it. As mentioned above, this avoids the "this is not my data" reaction when they see a spaghetti diagram, and KNIME functionalities like workflow annotations, components, and node annotations ("self-documenting") helped greatly with communicating the transformations and reasoning behind them.

16

Creating a Reference Process

Creating the data model for your process mining efforts is just (more than) half the battle. There is one more preparation step that you have to do before you get to the analysis in your process mining tool, and that is the creation of a reference process—the formal description of how work should be done so that you can compare it to the data set to see how the work was really done. There might be some discrepancy in the perception of the managers, and this typically is an eye-opener for them (and your auditors as well).

But creating a process model also opens the opportunity for a first analysis, which was the analysis method for decades.

This chapter will explain all the activities for creating a reference process using the example of the Heartbeat process and then will extend this to creating future-state process models and the simulation of these.[45]

Import of client-provided process into modeling tool

The first step is to either enhance the napkin-level sketch of the process in scope to a formal BPMN diagram by interviewing SMEs or facilitating process workshops, or—as in our case here—importing the provided *.bpmn file into our process modeling tool. And as expected when using a standard file format, the process imported without any issues. Unfortunately, that might not be the case all the time because some vendors apply some "interpretation" of the standard to their implementation and add tags into the code that you will have to remove before the import.[46]

There is also the option to export a discovered process in Process Mining into the modeling tool if you don't have any existing diagram. In this case, I like to go into the mining tool and select all variants that cover 80% of all cases (an arbitrary number, but typically, you will see that the first 5–10 variants cover the 80%, and then the variation curve goes into the "long tail" phase of one-off process instances). This can also be a starting point for your process workshops.

In our case, we had an existing client diagram, and Figure 16.1 below shows the imported process on top of the graphic.

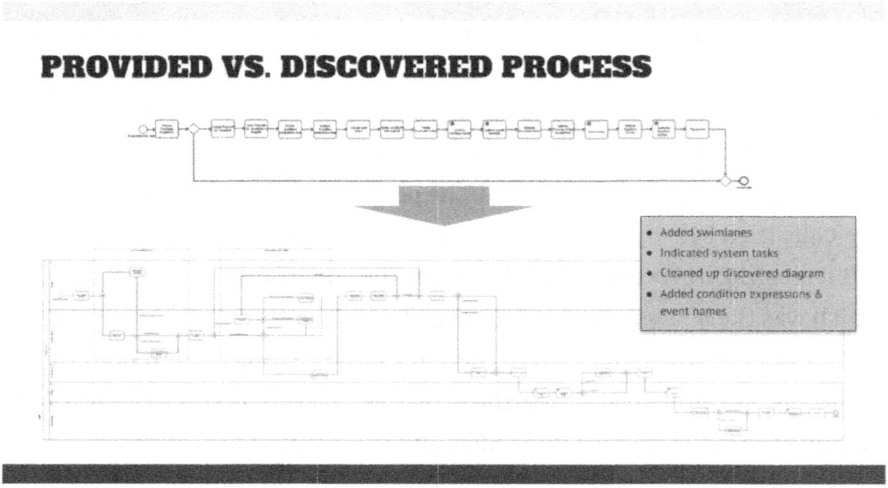

Fig. 16.1: Imported client process and enhancement

However, the imported diagram might tell you "only half of the story," and when you take the information from the data modeling activity, you can add more information to it:

- The biggest visual change is the **addition of swimlanes**. Typically, you use those for roles or applications, and the data set includes the user information. Once you have created those, add the events, tasks, and process logic into the lanes.

What you might realize is that the process now becomes seemingly convoluted and complex. This is an indication that the process might not be as streamlined as the process owner told you at the beginning of the project.

- The second step is to **indicate the automated steps** (indicated by the "system" users). In our scenario, you might want to take notice that we have internal system users and external system users, where Heartbeat has given vendor applications access to perform steps in the process. This is not very common, and typically, you will model external interaction as separate pools with message exchanges, but I am not aware of any process mining tool that understands the concept of pools, so you will either add the external steps as if they were part of the internal process, or you will place them in another swimlane. In our example diagram, they are modeled in-line because the client did not distinguish between internal and external automation (but you can still analyze that later by filtering for the different user types).
- The third activity was to **clean up the diagram** and try to get the most understandable sequence flow possible by arranging the lanes and other objects. If you run into trouble doing this, you might have a first indication of sloppy thinking and a not-so-great process design. Moreover, keep in mind that many diagrams created by business users are not living up to the standard that one would expect from a trained business analyst. The diagrams might not follow the BPMN standard, and you will have to add steps and other things (one item that I see frequently is that gateways are considered "decision points"—aka activities when someone decides. For this, you need to use a task in BPMN because the gateway simply routes the process flow once the decision has been made). Please ensure that you explain this when you review your cleaned-up diagram with the process owner.
- You also want to clarify the **condition expressions** when you branch

in a process. This information was not included in the provided client diagram, so you need to discuss this with your process SMEs. This information will give you an initial set of business rules, such as "over $50" and "below $50" as thresholds for when to choose a path after a gateway.
While you are at it, ask the SMEs when this threshold was defined. Most likely they don't know this, and I have seen many organizations where the business rules were defined decades ago. This raises the question if it might be too low and therefore creates unnecessary complexity and approval steps in the process ($50 in 1980 is the equivalent of $195 in 2025).

After enhancing our process diagram, we see that there are four distinct phases in our process:

- The purchase requisition
- The quotation and order creation steps
- The order fulfillment
- The payment processing steps.

We also see that there is quite some back-and-forth going on between roles in the first two phases of the process, while the last two seem to be relatively straightforward.

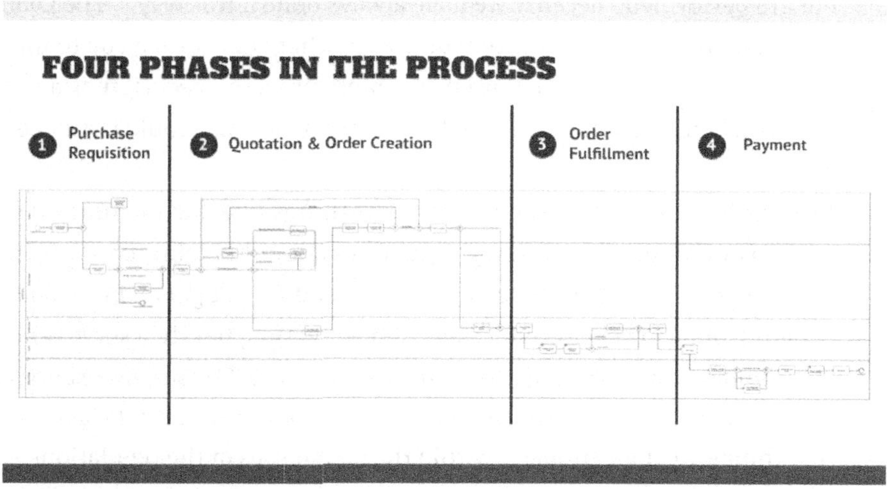

Fig. 16.2: Determination of the different process phases

Base process analysis

After you have cleaned up the process model, you might want to apply some out-of-the-box base analysis that ships with your tool.[47] These are typically reports for calculating the number of handovers between roles (fewer are better) or the use of applications and the switch between them (also, fewer switches are better).

The idea is to find what the Lean folks call "value" and "waste":

- **Value**—the measurable business outcome of a process. This can be the ultimate outcome for a customer ("We solved a problem," for example) or be a sub-outcome in a longer series of processes that ultimately will end in a customer outcome. An example of that could be the creation of a purchase order in a procure-to-pay process, which is needed to complete the first step of that process and trigger a vendor to send a product or perform a service.
- **Waste**—this includes all activities that do not contribute to value creation

but are performed "because we have always done it this way." The Lean folks have a very tight definition of waste[48] that I encourage you to take a look at. Examples of that are incomplete process execution or a low quality of the outcome, which will require rework that could have been avoided.

Please note that this does not mean that things like documentation of systems or things required by regulators should be eliminated from your processes just because an end customer does not get an immediate benefit from it. Those activities are value-creating, but they serve other stakeholders than a paying end customer—a future IT team, for example, that is tasked with updating a system. Or the public that can be assured that things are done properly within the boundaries of the regulations in the two examples above.

Fig. 16.3: Types of waste in Lean

At this point in the game, you might not know all of this, but you might want to discuss this with your SMEs in your process workshops and take notes.

Typically, a person who executes the process knows where things are stuck or need improvement, and you want to add this to your hypotheses and see if the "gut feelings" are true based on the data set.

When speaking with your SMEs, also try to gather additional contextual information about the process. Also use the analysis reports in your modeling tool (reports, queries) to perform a base analysis.

This might include the following areas:

- **Process**: Number of processes/tasks, organizational changes, tool changes, internal and external handovers, approvals and checkpoints, pain points, and process gaps.
- **Applications**: Number of tools, tool redundancies, unique tool combinations, technical interfaces, and workflows/integration.
- **Organization**: Number of org units/roles, consolidation of redundancies due to process improvements.
- **Data**: Project deliverables, messages on handovers, and overlap in data/deliverables during the process.
- **Risk and Controls**: Number of identified risks and controls, percentage of risks mapped to control framework, degree of risk mitigation with identified controls.

Lastly, you might want to revisit the areas where you struggled to clean up the imported process model and discuss this with your SMEs as well. In our example, shown in figure 16.3 below, there are two approvals needed (one for the purchase requisition and one for the quotation), and they are handled inconsistently—this could be a potential compliance issue.

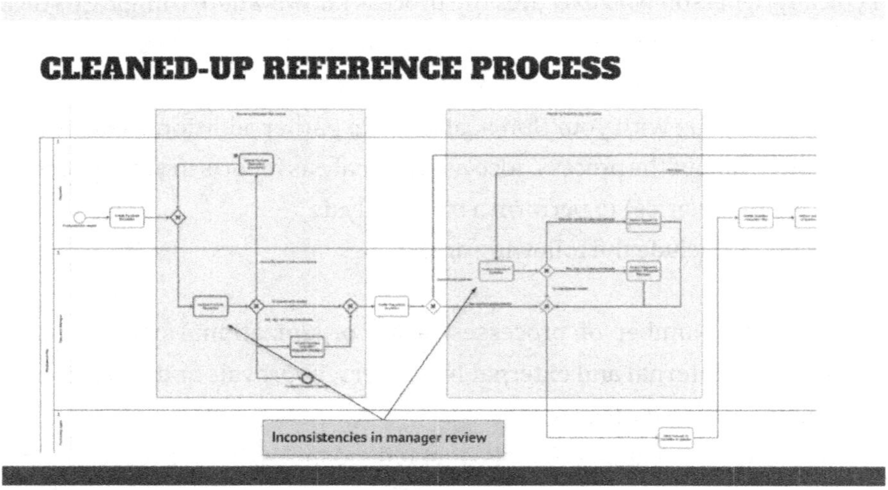

Fig. 16.4: Inconsistent process pattern with potential compliance issues

The result of all of these activities is two-fold:

- You get a clear understanding of the as-is process, and
- You might already have found some areas for improvement.

Creating the future-state process

The next step is something that you might have to do in multiple iterations depending on the additional analysis that you will do in step 5 of the six-step approach later: you want to create an "enhanced reference process" that includes the improvements from your workshops and your suggested ideas in a separate diagram (see Fig. 16.5 below that shows the improved order quotation approval steps).

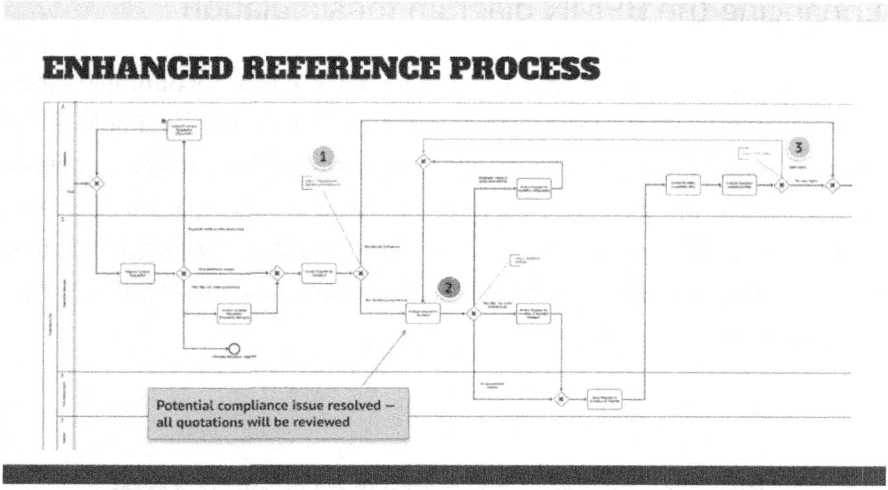

Fig. 16.5: Enhanced reference process

This allows you to do the conformance check twice in steps 5 and 6 of the six-step approach:

- How work should be done when you started the project (the cleaned-up model that has no improvements) and how compliant the execution was compared to this process—as part of step 5.
- How work should be done going forward. With that, you can determine if your conformance will improve in step 6 later.

We will talk about conformance analysis in the next chapter in more detail. Just keep in mind that you will have to upload the BPMN diagrams to your process mining tool and then map the activities of your event log to the steps in the diagrams.

The steps for this are different from tool to tool, but in the ARIS platform this is solved very straightforwardly by simply pressing a button to send the process to Process Mining and then doing the mapping of steps in your data model.

Enhancing the BPMN diagram for simulation

One process analysis method that you should use in process mining projects is simulation.[49] There are different types of simulations available in the different tools, and my suggestion is to have a simulation tool that is outside your process mining tool but can use data that was collected and calculated in the process mining tool—depending on your vendor, this might be almost fully automated or might need some "swivel chair" data entry. Either way, it should not take too long.

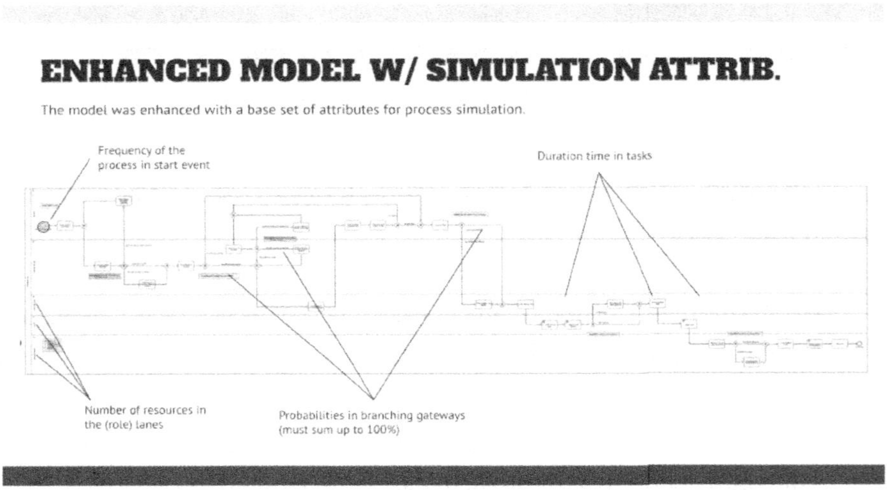

Fig. 16.6: Additional information in diagram for process simulation

When preparing your simulation model, here are a few attributes that, I think, are the minimum information that you should add to get a meaningful result:[50]

- **Frequency** information in the start event(s). How often does this process occur? Daily, 10 times a day, 50 times an hour, 10,000 a month?
- **Task duration**: here you can go into a very detailed setup. Imagine you

have a production process—you require orientation time to prepare a machine, you have the actual processing time for the step, and then you have wait time (for example, to let paint dry). My recommendation for your first simulation is to add the average processing time as the only number.

- **Probabilities** when you branch. You need to tell the simulation engine how many processes shall go down which path. When entering that number, please verify that it adds up to 100%.
- **Resources**: The majority will be roles/humans, but you can also enter technical and other resources that come with their set of attributes that might be relevant. For roles, I suggest the following information: number of employees for that role, cost for the busy time, and cost for the idle time (which for office workers most likely is the same number). Enter this as "per hour"—the simulation will calculate not only the utilization of the roles in percent (so that you see if you have enough employees) but also the cost for the time that people are actively working on the tasks. For technical resources, I recommend adding them as a separate swim-lane and adding the cost as "Average total cost" to the tasks, even though you could also go very deep into the details on these resources as well. What the simulation engine will do is to calculate the cost by multiplying the cost per task by the number of activations (every time the simulation starts this step).
- **Schedule**—by default, the simulation engine will do its thing, assuming all 24 hours of the day are available. That obviously is not the reality in organizations, and therefore you can configure a schedule, which can be as simple as an "8-hour day" (recommended when you get started) but can be as complex as a schedule aligned to a real calendar (to capture weekends and holidays) for each role. When you have part-time roles (think: retail salespeople), you might have to change your process model and add branches to accommodate the different numbers of full-time and part-time employees.

Simulation analysis

As I said before, you can get into very tiny details when configuring your simulation, and I recommend taking a look at this technology, but when you get started, keep it simple.

However, you can also run analysis on the simulation results—for each instance of the simulation and the comparison of multiple simulation runs after you change the underlying models.

For an **individual analysis**, you can compare the results to the SLAs and expectations that you have on the process execution, looking at the criteria below, but you can also generate additional charts that will show you bottlenecks (by plotting out processing time and dynamic wait time for each step in a bar chart) that you don't necessarily see in the results of the full process. This can give you some ideas of how you might want to change the process model, for example, by increasing the number of employees in a role and seeing if that removes the bottleneck.

For obvious reasons, this is a very process-specific analysis that will change from process to process that you simulate so that I cannot cover all the details here. But the documentation of your simulation tool might give you suggestions of what to look for (hint: it is most likely a subset of the things we will do in the next step of our six-step approach).

For the **overall analysis**, you might want to create a baseline simulation for the initial process that you've modeled (the current state) and then run it again for the future state(s) that you come up with either in this step 4 of our process mining approach or in step 6 when you articulate your improvement opportunities and suggest a change in process.

Overall, I suggest capturing these metrics for each simulation run in a table and then comparing them to find the best result:

- **Capacity**—how many processes have been started in your simulation time frame? How many of those completed successfully were canceled/terminated or are still in process?
- **Time**—What is the average processing time for the process overall,

as well as the individual activities? Is there a difference between the simulations when you have a calendar-based schedule (machines don't do overtime and can run 24/7)?

- **Cost**—total cost of the process (human and technical resources combined). Also calculate the cost per activity in case you are thinking of automating some steps or larger parts of the process, for example, by using agentic AI.
- **Utilization**—degree of roles actively working on the activities versus the available time. This will help you to balance the number of employees per role, and some simulation tools have this calculation already built in and will give you a range of employees for each role for each simulated model.
- **Bottlenecks**—this is a qualitative measure, but note where the process showed these and how many you had.

Takeaways for the example project

Creating a formal process model of napkin-level process from Step 1, or importing and cleaning up the client-provided BPMN diagram, is more than just a formal step of preparing for the conformance analysis in Step 5 of our six-step approach.

We have already found some analysis results that can lead to improvement opportunities:

- The first insight is that the process is not as streamlined as the client might think. The fact that we struggled with cleaning up the BPMN diagram, especially around the two review/approval areas (purchase requisition and quotation), showed inconsistencies that might have regulatory compliance impact at worst or show opportunities for simplifying the process and rethinking the rules when a review should take place at best.

  Either way, changes in a future-state process seem to be warranted.
- The other aspect is that you got more conversations with the SMEs

and could capture additional insights from them while reviewing your process model and trying to understand the logic in it. These typically happen outside the "official" strategy meetings and might be more informal so that the SMEs might open up more, which is what you want. They might also acknowledge you as a person and not in your business role as the consultant or "process improvement guy who is after our jobs"—building that rapport will help when you do the larger analysis in the next step, and your findings might be more easily accepted and dealt with in a constructive way.

Capture the remarks that your SMEs make, even though they might be unrelated to the topic that you are currently discussing—maybe you detect a pattern that is hidden underneath the data that process mining visualizes. This could be process-oriented or non-process-oriented; such a lack of trust between the people involved in the department and management might take some action to improve that situation.

- You can do a base analysis using the analysis tools that your modeling tool has. This might lead to ideas for future changes. In our example, the high number of handovers pointed to the challenges in the first two phases of the process.

  On the other hand, the process seems to use applications effectively since there are only the two major systems involved, and they are aligned to the different phases within the process (number of application changes as the analysis for this).
- Simulation as an advanced analysis technique is valuable to enhance the understanding of the process by adding schedule and resource information, as well as cost information, to a process. Using the current-state process model for simulation gives you a good baseline for your proposed process changes in the future (depicted in future-state diagrams that then will be simulated separately so that you can compare the main process metrics and see if your proposals will make an impact).

17

Process Analysis

We already have some analysis results, but this chapter contains the main analysis that one would want from their process mining tool. However, all the different analysis types might be confusing for a new user.

The best way to do a structured analysis is to split it into three different parts:

- Get a good understanding of the process and the "lay of the land," the main characteristics
- Perform a high-level analysis
- Conduct more detailed analysis based on the hypotheses that you have captured in step 1 and the conversations that you already had.

As mentioned before, you also should have a place where you can capture your analysis results and ideas for findings. For this book, I have created a page in Notion with some findings and ideas that you might want to check out as an example.

The detailed findings should be collected on a shared page (for example, in Notion, as shown in figure 17.1 below) so that everyone can contribute. This page should also be used to collect the improvement ideas in preparation for the prioritization and rating of them in the next step of our six-step process.

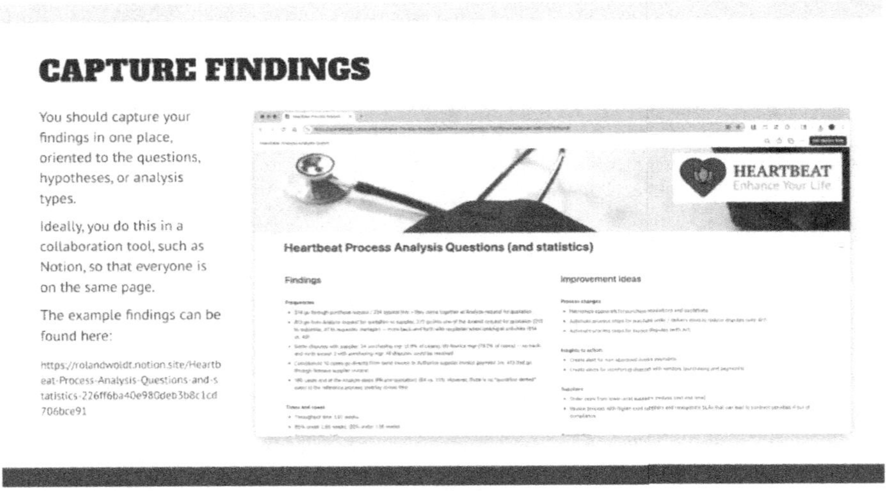

Fig. 17.1: Notion page to collect findings and ideas

Another good idea is to create a Fishbone (cause/effect) diagram[51] that is aligned to the questions that your client might have. Either way, you should follow a structure for your analysis and not just "poke around." Without focus and a structured approach, the probability of getting to meaningful findings and improvement ideas is low.

And also keep in mind that this step, just like the others before, is iterative. You might go back to a different analysis and dig deeper to find results that might answer a question that you find elsewhere.

To keep the project on time (remember: you want to deliver value in short 4-5 week projects), I recommend limiting the number of iterations. There is always the option to put things on the "parking lot" list and tackle them in the next iteration of the project.

Understanding the process in scope

As we've seen in the previous chapter, our process is comprised of four distinct phases (purchase requisition, quotation & order creation, order fulfillment, and payment; see figure 16.2), but there are different aspects you want to understand in addition to this:

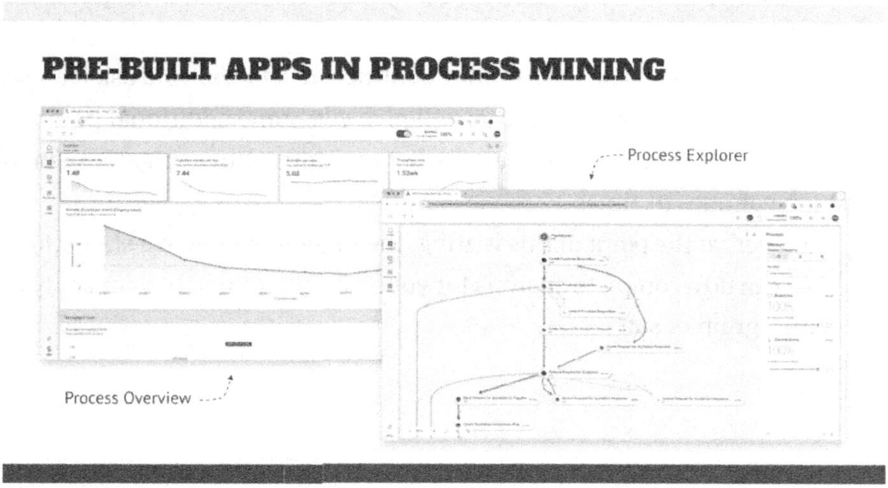

Fig. 17.2: Standard "apps" in process mining tools

- What are the **main characteristics of the process**? For this, your tool might have a standardized "Process Overview" dashboard that shows you the numbers of the cases, for example, how many cases happened in the data time frame, how they are distributed over time, etc.
  This gives you a good idea of the "what," and I recommend creating a standardized version of this if your tool doesn't have a pre-built app for this.
- The second pre-built visualization in the tool is a "**Process Explorer,**" which became the infamous picture of any process mining tool. At this point, simply use it to get a "feel" for the process when adjusting the controls for the number of activities and number of connections ... and yes, if you move them to the right, there is a high probability that you will see the dreaded "spaghetti diagram" ...
  What you want to understand is how complex the activities are in each

of the four phases of our process and how many loops are in the process. This will give you an initial idea of what to focus on in more detail when you have to answer the questions that your client has put in scope for the project.

- And then you can use the **AI functionalities** in your tool and ask it about anomalies in the process. This will give you some additional pointers that you might want to put on the list of things to investigate, such as high volumes or times.

However, at the point of this writing, the AI tools are not at a state where they can do a complete analysis for you. Take their input to your analysis with a grain of salt.

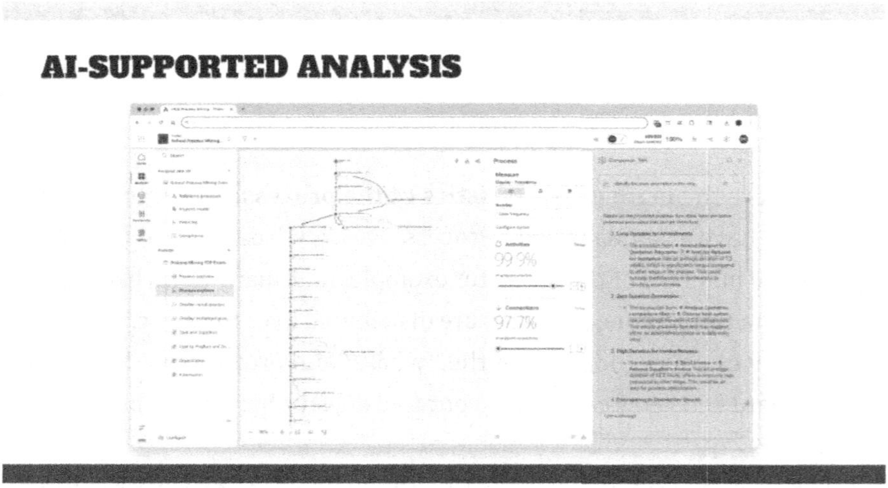

Fig. 17.3: AI-supported analysis features in process mining tools

High-level process analysis

Let's get started on taking a closer look at the different analysis techniques that are typically built into process mining tools, what they do, what you can get from them, and what typical next analysis steps might be.

The objective of the high-level process analysis is to discover the main findings or trends, even if the time for more detailed analysis might not be available. Of course it is better to be thorough, but sometimes things need to happen to meet a deadline...

Figure 17.4 below shows the activities for the high-level analysis in an overview—as you see, the analysis types will require multiple tools and not just your process mining tool alone...

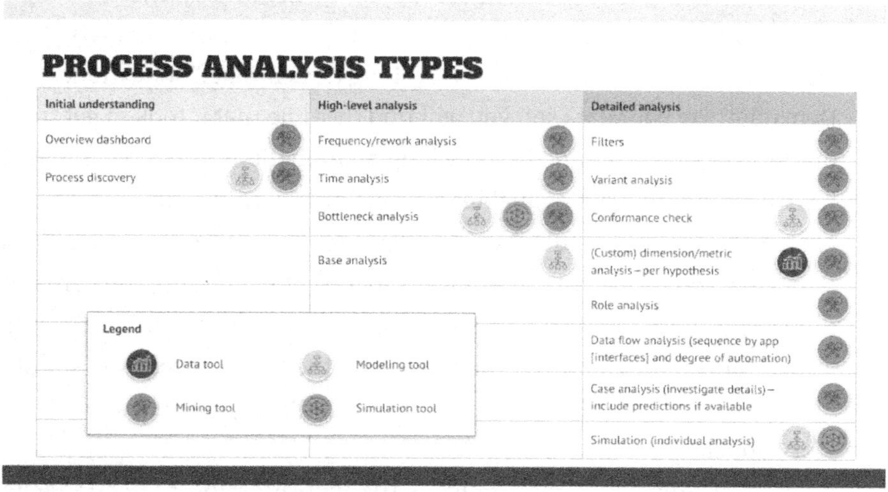

Fig. 17.4: Process analysis types in an overview

- The first analysis type that you want to look at is the **frequency analysis**. You do that in the Process Explorer view that is shown in figure 17.2, and it answers you the question "How often does this happen?"

What you see is the frequency of cases that go through activities and through the different paths of your discovered process. This will show you the "happy path" and loops that might or might not be wanted (in the process model that we created in the previous chapter, these show as the different paths after a gateway). The interesting thing is to see if there are certain patterns or if there are paths taken that might not adhere to policies, such as the payment of an invoice without approval (compliance issue).

In the tools, you can switch from the case numbers to an activity view. Here you can see how often the activity has been triggered—if that number is greater than the number of cases that went through that activity, then rework has been performed. For obvious reasons you don't want this, and you should note where rework happens so that you can further investigate the reasons for this in the detailed analysis.

- The next analysis is done on the same screen—the **time analysis**. This will give you the answer to the "How long does it take?" question.

  Depending on your data set, you will find the time that it took to get from one step to the next and also how long an individual step took. You will have different metrics available: average, mean, minimum, maximum, etc. These can be useful when you want to do 2- or 3-factor analysis later (in our data set, you might want to look at which suppliers are not only the priciest but also which ones are the slowest and see if there is a correlation, for example). Just like with the frequencies, you should identify and note the longest activities and paths and see if the start of a story appears.
- Related to times and frequencies is the identification of **bottlenecks**. Some process mining tools have animation features that allow you to "play back" the process. They show you the individual cases going through the discovered process model and "wait" at steps per the timestamps in the data set. This gives you a visual impression of where the process slows down or comes to a halt (and it has a meditative effect just watching the small bubbles go through the maze).

  A better way to identify bottlenecks, though, is to use simulation. I described simulation in the previous chapter already, but you can use the metrics from process mining and add them to your simulation model and run a baseline simulation. The tool will create charts that show the bottlenecks clearly, and you can overlay them on your process diagram and highlight where problems occur.

- And lastly, you have the results of the **base analysis** that was described in the previous chapter. The results of the analysis based on the process diagram include information about application changes or the degree of handovers between roles (both should be minimized if possible) and might give you some ideas about where to do changes in the future.

Figure 17.5 shows the process diagram with findings highlighted on the process map. This visualization can be very useful in addition to your data collection in Notion.

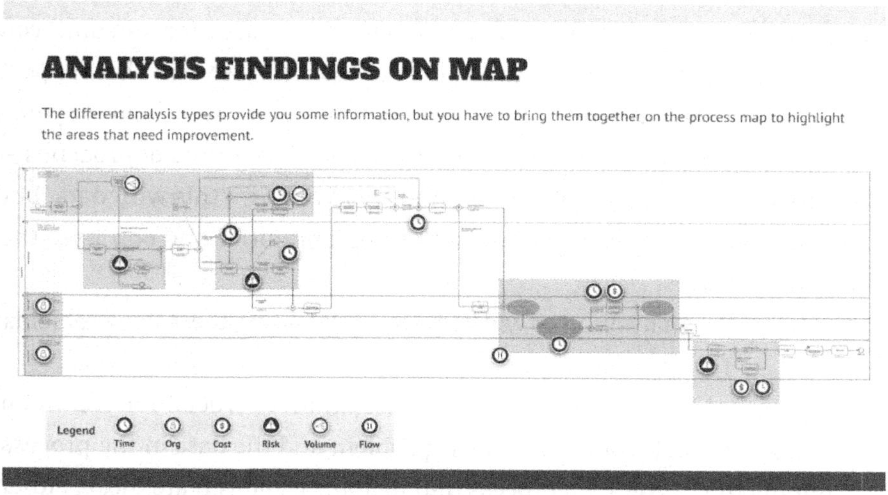

Fig. 17.5: Analysis findings on the process map

Detailed process analysis

We are reaching the state now where we will take a closer look at the hypotheses that we figured out in step 1 of our six-step approach—for example, if SLAs are met or if the organizational setup is appropriate for the volume and execution of the process. The objective here is twofold:

- On the one hand, you want to **confirm or reject the hypothesis** while looking at different aspects such as where the process runs well and where roadblocks occur (the "what").
- On the other hand—and that appears to be more important to me, but I have seen "forgotten" over and over again—you want to **figure out the reasons for the undesired process performance**. And you might find that the reasons for these might not be process-related at all but rather have their origin in contracts, physical environments, regulations, etc.

Here it's key that you will have to ask "Why?" over and over again until you get to the root of the problem. And all of your tools are just tools that support you with this; they will not raise these questions nor answer them for you.

This is why I am a secret admirer of the Lean/Six Sigma approaches—they are looking for the right things and try to confirm this with data, but unfortunately sometimes get lost in the small minutiae and don't see the forest for the trees.

The following analysis types will help you with solving these two problems:

- The first one seems to be a tool feature, but it is crucial for you when you want to tell your story of the problem and the data in the process mining tool: **filters**. All process mining tools let you create those, either by selecting a measure in a dashboard, an activity, or a connection in Explorer, and the tool will select only the data that is true for this selection.
  When using filters, establish a naming convention that will help you remember in which context you want to use your filter. "Cases by org unit

A" is not as good as "Escalations—software incidents—org A" when it comes to telling the story of how technical incidents in your organization work well for other types but not for software. The names should be aligned to your hypothesis in focus.

- The **variant analysis** is the next analysis type you want to look at. What is the distribution of the cases in your data set? Can you see if the majority of cases follow a few variations, or do you have tons of different variants and you cannot find any patterns?
  What I have seen in mining projects is that 80% of all cases typically are in the first 5-10 variants, and then it tapers down into the "long tail." For obvious reasons, you want to have a low number of variants so that you can see that your people follow a process and not just make up stuff as they go. But you also want to take a closer look at each variant: what is the process flow in them, and how often does this variant occur? How long did it take? Is there a second or third dimension that might have an influence—i.e., does this variant just happen in one part of your organization, and why?

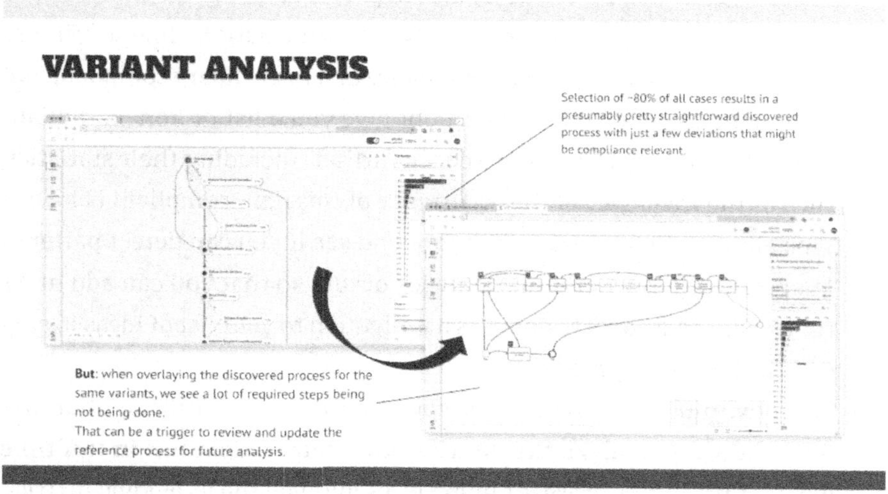

Fig. 17.6: Variant analysis and check of missing steps (compliance issue?)

- Speaking of following a process, one major use case is the **conformance check**: do my users do what they are supposed to do (and I have defined in a process model)?
  We spoke about this in the previous chapter, but when it comes to risk & compliance management and audits, the first question typically is if you have a process defined (the process diagram that is enhanced with risks and controls from your catalogs that are aligned to the regulations). The second question is if you are following that process in reality, and this is where process mining shines. You can upload the BPMN diagram of your designed process into the tools, and they compare the data with the prescribed process and not only give you a statistic to which degree you are compliant (even over time if you wish), but also give you hints where you are not matching your prescribed process. This can cover wrong start or end events or the wrong sequence of steps (the invoice needs to be approved before payment!).
  In case of wrong start and end events, check if this can be a symptom of your data set: you might have a cut-off when your data set started, and you might have missed the first steps in your event log—in this case I recommend filtering out the affected cases. Or your event log might include cases that are currently still in progress.
  Some tools also provide you additional information, like a "fitness score" that calculates partial compliance and not a binary compliant/not compliant score, or they even might give you a list of non-compliant things that you might want to check further, including their statistics, such as the frequency of the occurrence of this non-compliant behavior. This allows you to investigate more and see if you can detect patterns or specific areas where this behavior occurs so that you can add more controls or additional training as a mitigation to your list of ideas in step 6 of the six-step approach.
- The next analysis type is very specific to your hypothesis, but you might want to **analyze the metrics and their correlation in multiple dimensions**. While process mining tools give you the individual metrics, some of them don't even allow a simple 2-dimensional analysis in the

form of a scatter plot (or 3-dimensional if you want to add bubbles for the third dimension).

To achieve this, you might have to go back to your data set and transform the data so that you can plot it out in one of KNIME's visualization nodes. In the example of figure 17.7, we see the scatter plot of the frequency of cases mapped to their duration. The added value now comes from the two axes that are added: where is the line for successfully completed cases, and what is an acceptable duration for the cases per your SLAs? This will give you four quadrants that you can improve with different methods as shown below.[52]

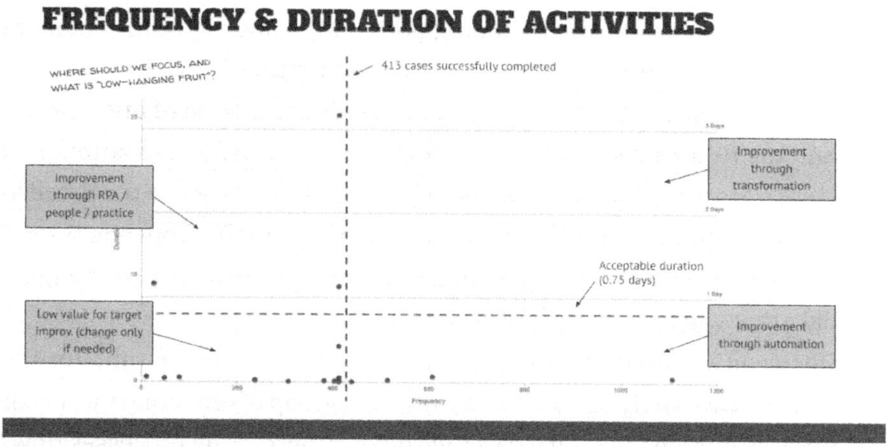

Fig. 17.7: Scatter plot of case frequency and duration

- Another analysis is to see **how users and org units interact** with each other (and if problems arise because people don't talk to each other). This can be accomplished by uploading your data into another project in your mining tool, but instead of using the activities as activities, choose the users, roles, or org units as activities. The resulting graph in the Explorer view will now give you the frequencies and the times for interaction within the process.

This allows you to identify the most "popular" and "unpopular" users or roles, and that might give you an idea of what to expect when you do the **utilization calculation** of your users or roles within the process. This calculation can be done in your data or your simulation tool, and in the latter you can also make changes and re-run the simulation to see what the impact of more employees on the process will be (as part of your opportunity validation in the next chapter).

Another organization-relevant analysis is the **cost analysis of the roles and users** that you might want to have. Simulation can give you the calculation of "busy time" versus "idle time" and calculate the "busy" cost and the total cost, which also includes the cost for technical resources or machines, while also including preparation and wait times that are needed for the process (and you pay people for them as well).

- The next topic you might want to look at is the degree of **automation and the flow of data**. What is the relationship of manual to automated steps? Are systems used in silos and not integrated (or use the dreaded "swivel chair interface," in which a human enters data from one system into another—a massive opportunity for making mistakes and changing data that will lead to errors downstream)?
- Once you've found the thing you looked for, it might be time for the detailed **case analysis**. Look at the flow of a group of cases that meet your criteria and the individual cases within it. How did the processes flow? Did they take the same amount of time or paths? Were there escalations or other anomalies?

  Some tools also use machine learning to make a prediction of still-running cases and predict (with "x% probability") what path this case will take and when it will finish. This information then can also be used to trigger alerts—the "insights to action" functionality that we already mentioned elsewhere can be applied here as well. When the case is in danger of missing a threshold, you can be informed and interact with it before the event even happens.[53]

- Lastly, you can use the **simulation** tool for analyzing the process as well as described before, and it will give you statistics for cost, utilization, and so on that you can export and use in a dedicated data tool for further analysis. It also will show you the bottlenecks of the process in a visual format that you can then lay on top of the process diagram.

  But the real "magic" of simulation lies in its prediction capabilities that we will discuss in the improvement opportunities chapter.

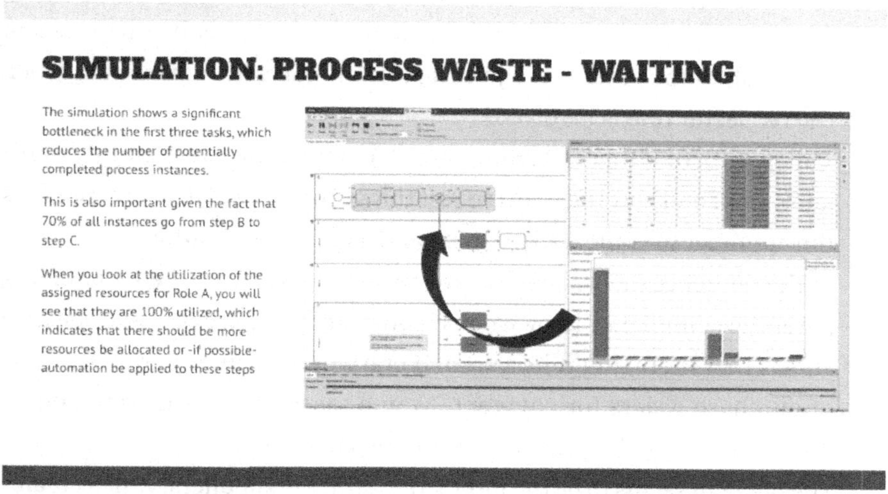

Fig. 17.8: Simulation result and visualization

As you have seen, doing your process analysis in these three steps will add more granularity to the results and makes sense depending on the available time you have on your hands for the project.

The critical point in all of your analysis is not only to show the "what" ("Look how bad this process runs") but to ask the "why" question until you have found the reasons for why the process is so bad. For this, a process mining tool is a good tool because it is data-driven, but you must combine it with other analyses and tools, as well as speak with the subject-matter experts to get the context of what you find. This will ensure that your identified opportunities in the last step of our approach are grounded in their experience and will not be put aside as "ivory tower ideas."

Takeaways for the example project

The example data set and the analysis that was done in the four tools showed many areas that could need some improvement:

- When we look at the process itself, we see a few areas that take a long time and have volume, such as the back-and-forth when it comes to approvals of the purchase requisitions or the quotations, especially when the requester role is involved.
  This points to either a training problem (the requesters simply don't know what and how they can request things) or a lack of data that requires the approver to come back to them and ask for clarification.
- In the reference model analysis, we've already seen that there might be a potential compliance issue when it comes to the approvals, and the data shows that there might be a second one when it comes to the approval of purchase orders for payment—some cases bypass this step and go directly from sending the invoice to the payment.
- The big item seems to be the interaction with the suppliers, which creates a long processing time and the need to settle disputes with them—by the purchasing agent and the financial manager. It is also interesting to see that there is a correlation between the higher-priced suppliers, the number of disputes, and the cost that these create. This points to the need for reviewing and potentially renegotiating SLAs with them or to rethink the procurement strategy overall (read: shift volume from the high-price suppliers to the lower-price suppliers).
- Using simulation, this situation and the bottleneck when creating purchase requests became obvious.
- When looking at cost, the supplier situation stuck out, and differences

between the average higher-priced supplier and the lower-priced suppliers were significant.

The distribution of products was roughly 25%, 20%, and 55%, and the average case cost spanned from $165 for office supplies to $215 for electronic parts, while potential savings by simply shifting to lower-priced suppliers not only reduced the cost from the average by 14-19%, but this will also improve the time by 36% (compared to the average case time).

Overall, though, the correlation between the number of orders and case cost showed, which means that we don't have any "special event" happening that would break this correlation. However, this also means that the problems are systemic and part of the process execution.

- There was also a relatively large number of variants (98), but the fact that 19 of the variants covered 80% of all cases indicates that there is a lot of manual (re-) work and potentially a lack of process knowledge or lack of consistent data in the process.

  This is supported by the fact that on average, the tasks done by the financial manager are 130% of the cases, while the number of tasks by the purchasing agent are 254% of the number of cases ... 154% more than if everything would go as planned. Interestingly enough, there are no disputes to be settled by the purchasing agents in the 80% group of cases. They seem to do a good job in settling the conditions with the suppliers before a purchase order is cut.
- The automation rate is relatively low, with 18% of all activities (including the steps that are done in the connected supplier systems), so improving this rate will not only help with data consistency and workflow management but also improve the times and reduce the number of reworks.

There were also SLAs defined that should be checked by the project:

- The duration between "Create Purchase Requisition" and "Analyze Purchase Requisition" should not exceed 1 day. The process mining analysis shows an average of 17.3 hours.

- Duration between "Analyze Request for Quotation" and "Create Purchase Order" should not exceed 7 days. The process mining analysis shows an average of 1.84 weeks, and therefore the SLA was missed.
- The financial manager role should just spend only about 1% of their time in this process—they are busy with other stuff. The calculation of utilization shows an average utilization of 4.55%, and therefore the SLA was missed.[54]

Speaking of utilization, the analysis also showed a significant over-utilization of the purchasing agent role (~145% utilization), which also contributes to the long times, for example, when settling disputes with suppliers. However, since disputes with the suppliers are also showing up when it comes to the invoice payment (with mostly the same companies), it appears that there needs to be a clarification of the collaboration between Heartbeat and those vendors warranted.

18

Sidebar: Dashboards and Storytelling

This chapter feels a bit "off" in this segment because it does not justify its step in the six-step process but is part of the presentation of the process analysis done in step 5. However, I have seen too many "bad" dashboards that missed the point in communicating the correct message, and the fact that the process mining tools are built to create dashboards makes me think this extra chapter is warranted.

Dashboards versus reports or presentations

We are all used to listening to presentations in our day-to-day businesses, and even our kids learn how to use presentation software in school (and, to my dismay, also the preferred way of "presenting": bullet points).

This has some serious disadvantages when it comes to process mining:

- Presentations or written reports are a snapshot of findings and recommendations at a given point. They don't update and most likely will be forgotten before the presenter is done with going through his or her slide deck—this is commonly known as "death by PowerPoint."[55]
- The bullet format seduces the author to do a "brain dump" of things that they found, and the mass of data might not lead to the distillation of information, or even better knowledge. The "best facts" will be listed,

and the most important questions (Why? Who? How? Where? When?) are not asked.

- The audience of the presentation zones out, or to say it in business terms, "starts to multitask," and any content will be heard but not understood or retained. Maybe after the deck is sent around, the recipients use an AI to summarize the content in a handful of bullets, but all the context and detailed information are most likely lost.
- Since the presentation format is so widely used, the presenter does not tell a story but lists facts and does not explain how things might evolve.

Dashboards on the opposite side have tons of advantages over the report format:

- They are always up-to-date if you connected a source to your process mining tool. This will show changes in the KPI over time, and you can see if the benefits that you proposed were realized.
- You can drill down into details in a dashboard and filter or show more details dynamically. This might lead to more insights and user engagement.

However, there are also some things to keep in mind when it comes to dashboard creation, and those are the topics of this chapter:

- Dashboards might be badly designed and not tell a story.
- The extra step of creating dashboards for the stakeholders versus the dashboards that are created for analysis purposes is not done.
- Dashboards and their widgets are not structured well, or the "wrong" widgets are used.
- The widgets are not formatted the best they can (or the tools don't allow this), which makes it harder to understand for the user.

But to be clear, sometimes it makes sense to write a report and spend the effort to assemble a long-form analysis document.[56] But a report is not meant to be presented where the presenter is the key person. It is meant to be sent/published and read in a focused session without distractions. They also should aim to be "timeless" and bring insights even if the report was developed months or years ago.

Dashboard setup

The first question that I hear in a process mining project from my younger colleagues is "What dashboards should I build?" That question is easier asked than answered, but here are some thoughts that might help you to decide:

- The first thing is to think about **who the audience** for your dashboard is. When you get started, it is not the stakeholder, but it is yourself. Think about it: you don't know what you will find, so there is some investigation and discovery to do.
- Then you might have some **standards already defined in your organization** about what needs to be shown in any case. Some tools have out-of-the-box dashboards already configured that the organization has adapted as the standard (figure 18.1 below shows the "Process Overview" that shows case information and gives a first impression on the volume and the performance over time of the process in scope).

PROCESS OVERVIEW

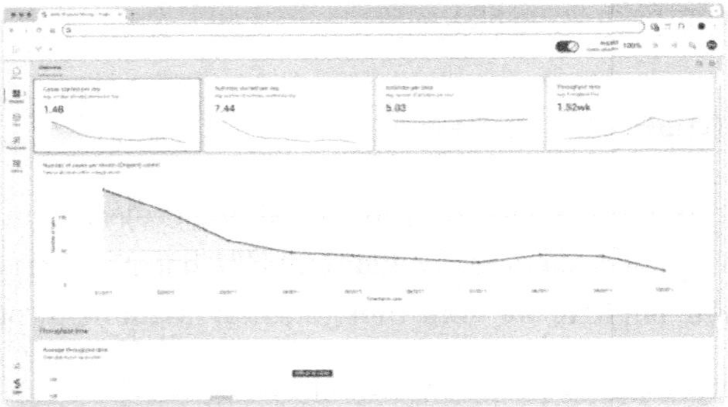

Fig. 18.1: Process overview dashboard

- The next thing that you should **look at are the hypotheses** that you've identified with the client in the previous steps—especially Steps 1 and 4. You want to answer all the questions that might have come up, and typically, you take each subject and will create a dashboard for it. Figure 13.3 in a previous chapter shows some examples of questions that you might encounter.
- Keep your dashboards simple: **one dashboard for one idea** (or area of research). Do not try to be smart and create dashboards with tons of filtering options and sections for different use cases or hypotheses. One answer per dashboard to one question is plenty, and your audience will thank you for the clarity.
- Sometimes you will see that one dashboard is not enough to answer a question/validate a hypothesis. In these cases, **don't shy away from creating additional dashboards**. I suggest that you develop a naming convention for the dashboards that indicates in which area you are working—in our Heartbeat example, that could be "Organization—..." or "Automation—..." indicating each of the five areas that the client is

interested in.

- Once you have done your discovery/analysis dashboards, **think about how to communicate the results**. In most cases, I have seen teams omitting this step because "we don't have time for that left" or "I don't want to lose details." These teams lose the stakeholders because they have built the dashboard the way *they* think or how they have discovered their insights and opportunities.

Either way, don't be afraid to create more dashboards. In my experience, clients don't shy away from volume as long as they understand the structure and can "hone in" on what they are especially interested in. Having a clear naming convention in place and being consistent over multiple projects brings you a long way.

Dashboards structure and story

In the previous section, we discussed that there is a difference between the "work" dashboard and the "presentation" dashboard that is meant for your stakeholders. Here we want to take a closer look at the second type of dashboards.

The first thought is to go through your work project and determine if you can show all the sausage-making to someone who is not interested in the details of how you came to the conclusions or who needs to use the dashboard for daily operations and monitoring of the process. This very likely requires a second set of dashboards, most likely in a second process mining project that uses the same data set.

In that additional project (or projects, depending on the number of stakeholder groups and purposes), you should determine what is relevant for them:

- In a **project context**, you most likely present to senior management, and they are a bit interested in what you did (a few slides should tell that story), but in any case, they want to know what the findings are and how they can change the process to make it run better, cheaper, etc. For this type of dashboard, it is OK having to drill down into an area to see the results.

- In the **business-as-usual/monitoring scenario** after an implementation, you want to show the main KPIs quickly so that the information is "in your face," relevant, and easy to understand.
- There might also be the need to **manage the process itself**, and you might want to create a "meta dashboard" that shows escalations, cases that might go South, etc. The target group for those is not the line person but the process management or process leadership group, who are thinking already about the upcoming iterations of improvements of the process.

In all cases, you need to think about the story that the dashboard shall tell and what actions the viewer shall take away. Is it an immediate action that one shall take? In that case, you would rather not have an overly complex dashboard but rather have less information and clear symbology established (think about if you would have to explain what that upper light on a traffic light means and why it has that cherry color, versus the clear instruction that the red light gives you when you approach the intersection ... hit the brakes!). Furthermore, keep in mind that the same symbol (the red light here) might have a different meaning in a different context—be consistent in using symbols to reduce the learning effort for your viewers. If your organization does not have this in place, be the first to start a symbol library for dashboards, document your ideas/standards, and communicate this to everyone who is involved.

Once you have identified what dashboards you need, you want to focus on the individual dashboard itself:

- The first thing is what story you want to tell. In figure 18.2 you see an analysis dashboard that is structured to guide a user through it by first selecting the area of interest on the left and then gradually disclosing more information based on the sections in the right half of the screen. In that half, you see some general KPIs first (remember: you want to be able to ground things and be able to compare either based on selections within the dashboard or over multiple dashboards or even projects). Then you see two widgets that allow you to select more aspects that will be shown on the lowest level that holds the details in case someone needs that level of information.

By selecting something different, the structure still works, which—again—reduces the need for a user to "re-learn" the structure and creates the feeling of mastery and increases acceptance—try to have a similar structure of different dashboards to reinforce that aspect if you receive the feedback that "it works."

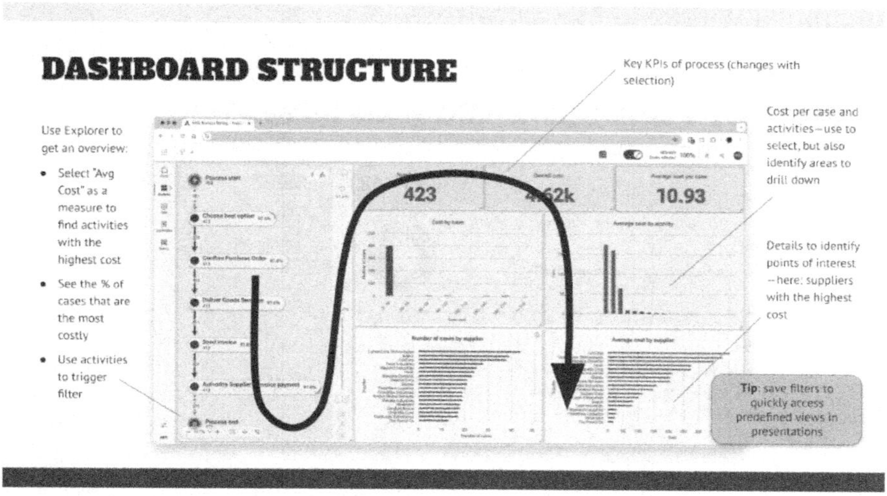

Fig. 18.2: Guiding a user through a dashboard

- Another example of this progressive disclosure of information is seen in figure 18.3—here the concept of "above and below the fold" is established. This comes from the time when people got their news from printed newspapers that were available at newspaper stands. To save space, they were folded, and the editors needed to make sure that the most important information was seen "above the fold" so that you would rather buy their newspaper than the competition's.

This same principle applies to our dashboards—don't "bury the lead"[57] somewhere down the page and force the user to scroll (they most likely won't do it). Rather, bring your main points to the prominent space on top of the page. Examples are the main relevant KPIs, selection opportunities, such as the product categories in our example below, as well as comparison charts that show differences based on the selection. Below the fold you have less essential information like the timeline chart or even the filterable table of the individual case details.

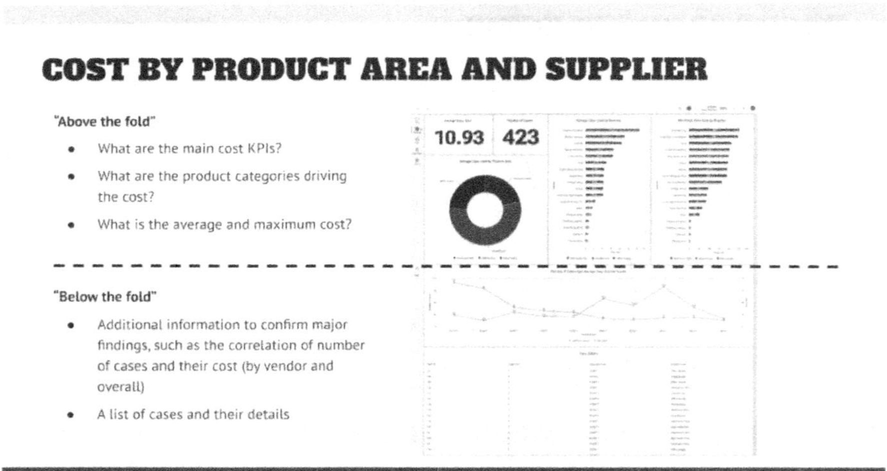

Fig. 18.3: Hierarchy of information

- Lastly, some tools allow adding headers, text, or graphics. Use these to create segments on the screen and provide information about the structure to even further guide your viewers. Use consistent graphics (symbology!) to show things like levels or progress if it adds to the clarity of your dashboard.

The takeaway for your dashboard is that you need to create them for your audience and what they need to know. Create a story that works for the dashboard, and determine an information hierarchy that will support your story.

Dashboard design

This is the section that most people think of when it comes to dashboarding—placing those shiny widgets on the canvas to show how smart one is. But unfortunately, I see visual illiteracy when it comes to that part as well, and the dashboards feel unprofessional at best and confusing at worst.

Selecting the right dashboard component

Let's take a look at which widget/chart type you want to use for which purpose, and I have to thank Mike Yi for the graphic in figure 18.4 below and the structure that he is implementing.[58]

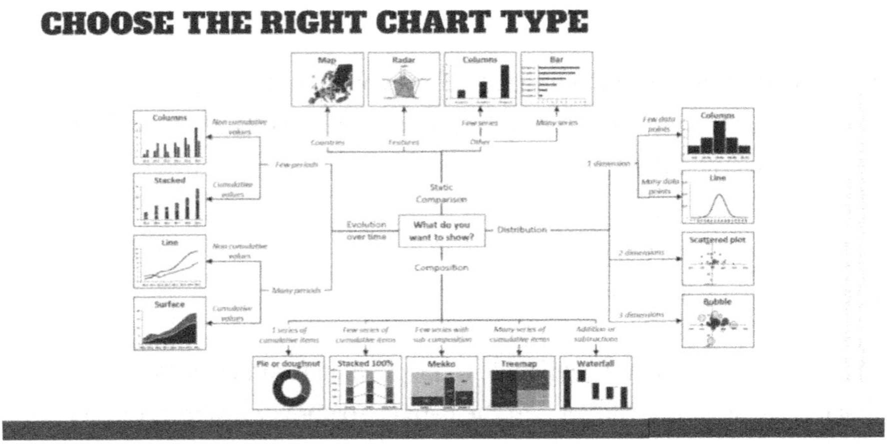

Fig. 18.4: Choosing the proper chart type

So ... how do you choose the appropriate widget type? Aren't they all the same? Well, obviously not. As figure 18.4 shows, you should check what you want to communicate:

- **Static comparison**: Display one piece of information/aspect and how it might relate to other information of the same type. This is the most used chart type, and depending on what you want to show (the content), you might select different types—maps, radar/spider diagrams, columns, or bar charts. Please note that there is a semantic difference between the last two, and the tools do not necessarily follow that semantic, and you can simply switch between column and bar charts by simply checking a box.
- **Distributions**: Show how information is distributed over multiple dimensions. The tools all show one-dimensional distributions in the form of column or line charts, for example, to show cases over time. However, the insights might be hidden in relationships between different information, and here you can (or cannot, depending on your tool) use scatter plots for two dimensions and bubble charts for three dimensions,

where the size of the bubble shows the third dimension.

- **Composition**: showing information in the context of the overall group of the available information. This can be as simple as a pie/donut chart or stacked column charts, but there are other visualization types, like mekko, treemap, or waterfall charts, depending on what you want to show—cumulative items or the decomposition of pieces of information compared to the overall group in waterfall charts.
  Just keep in mind that you don't use these types of charts to show absolute values. We humans are not good at comparing surfaces and deciding to what degree this represents the value. I like to challenge everyone to spot the difference between 29% and 30% in a donut chart.
- **Evolution over time**: Show changes over time, while keeping in mind the length of the time frame and if you want to show the individual values or show them in an accumulated way. For this, you will see column charts again (stacked or non-stacked), as well as line charts and their cumulative cousin, the surface chart.

Depending on the tool you use, there might be additional chart types available, but the message is that you should think about your message that you want to convey before just "throwing something on the dashboard."

Formatting of dashboard components

Lastly, you would like to tackle the formatting of the individual widgets. Take a look at the charts in figure 18.5 below; both show the same data, but the left one is using the standard formatting of the dashboarding tool, while the right one was cleaned up and has a clear message—the activities in the chart use roughly the same time, but four activities stick out and use progressively more time to execute. And no, you don't have to be able to read the activity names or any other text to get the main point of this chart.

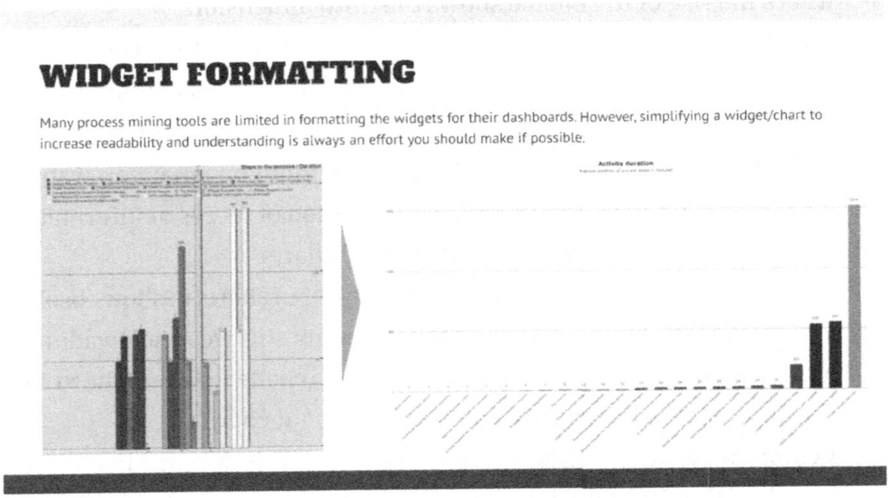

Fig. 18.5: Formatting of dashboard widgets

Here are a few tips when formatting your widgets, and unfortunately, most process mining tools are very limited when it comes to formatting (but they have pretty colors ;-):

- Keep scales consistent to not skew the data. In the example on the left, a logarithmic scale is applied, while the right chart shows the raw data. You can clearly see how the four activities have significantly higher values. Moreover, don't fall into the trap of just showing a range of data and cutting out areas where you don't see a lot of change. Think of charts of exchange rates of currencies or stock price charts that you see on TV or websites—they typically show just sections of the data, and differences now look significant, while the overall change of, say, an exchange rate of $1.10 to $1.11 is a difference, but relatively minor.
- Add titles and subtitles to explain what one sees on the chart. Some tools also allow adding tool tips to charts, and you should use this as your third level of information.
- Remove all unnecessary information, such as grid lines. Format them in a lighter color than standard (I am a fan of light gray) and add whitespace

where you can.

- When formatting the information, think of your story again and point out the things you want to show by applying a different color or, as shown above, by a sequence of colors that show the progression. All other, non-relevant information should be shown in a consistent, non-attention-grabbing color.
  Another example might be thresholds—do not color the whole areas in traffic light colors, but color the threshold lines accordingly, and add symbols for clarity. White space is your friend.

Conclusion

Dashboards are the best way to go forward if you intend your process mining result to have a longer lifespan than just the project read-out presentation. You might want to keep this checklist in mind:

- Build dashboards with your stakeholders in mind—in most cases that is not you, even though you might build dashboards for your analysis as well. Remember: "The worm needs to be tasty for the fish, not the fisherman." What do you want the users to do once they have viewed your dashboard?
  And occasionally the worm is a written document in your text editor that you send out via email instead of inviting everyone.
- Figure out what the information hierarchy of the things that you want to show on a dashboard is, and tell a story. Guide your user through the dashboard and give them plenty of opportunity to apply what they've learned. Repetition is good, and one dashboard per topic is plenty.
- Based on your story (comparison or change over time, for example), choose the right widget and format it in a way that takes away all unnecessary things. This became easier over the last 10 years when "flat design" became fashionable, but that pendulum is swinging in the other direction now. Resist the desire to add something because it "looks good" (like 3D effects that I, personally, despise). Rather stay simple and clean

and get your point across without your user having to think about what the takeaway is.

19

Identifying Improvement Opportunities

Well, we are approaching the last step of the project—the identification and justification of improvement opportunities.

But by now, it might be pretty clear where the opportunities are. If you have taken the findings of the various analysis types and laid them over the process map, it becomes pretty clear what "no-brainer" opportunities (aka quick wins with low effort) are, which opportunities are well-defined and need a bit of planning, and which opportunities require more research before they can be articulated and approved for implementation.

The process for this step is to create the initial list of recommendations, prioritize them with your stakeholders, create the business case for the important ones, and then get approval for the implementation.

And, of course, you can run a process mining analysis again once you have implemented the improvement opportunities to see if you got the benefits that you expected or if you have to make adjustments.

Initial list of recommendations

Based on your analysis findings, you should already have an idea how you can tackle the problems that you saw. However, I recommend aligning the improvement opportunities to the areas that you were tasked to research if possible when speaking with your stakeholders for the first time about this.

Figure 19.1 shows the list of ideas for Heartbeat in an overview.

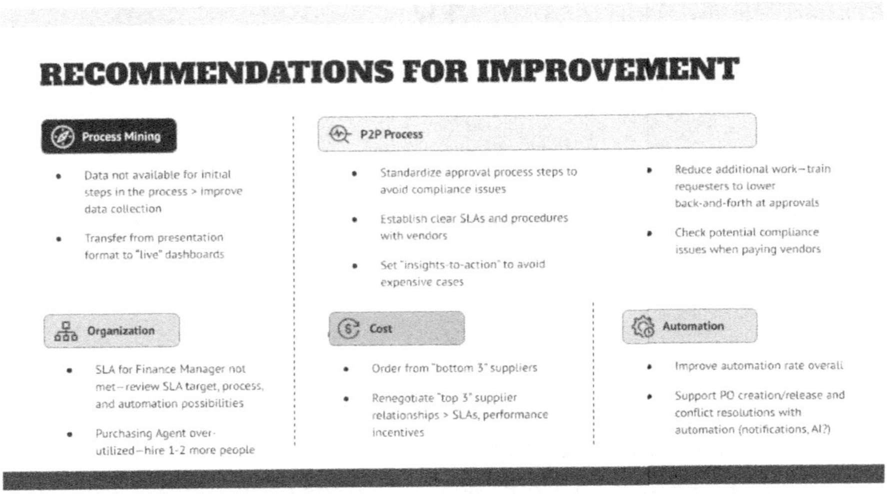

Fig. 19.1: Improvement opportunities aligned to questions

As you can see, these are "all over the place," so you want to create a list that sorts the ideas by two aspects: content and implementation effort. This will help you when creating a priority. Fig. 17.1 at the beginning of chapter 17 showed the Notion page that the project used for capturing the analysis results and details, and on the right-hand side you can already see the improvement ideas that were derived from that.

The key is to assess the implementation effort for these ideas:

1. You might have some ideas that are **easy to do and don't take much effort to implement**. Examples in our case are to configure thresholds in the process mining tool to trigger alerts when an interaction with the running process is needed, such as compliance violations (payments without approvals!) or when settlements are needed and the SLAs for these cases are in danger.
2. Improvement opportunities for which you **have a good understanding**

of the work that it takes already, but you have not created a business case for. An example here is the harmonization of approvals for purchase requisitions and quotations. This process change can easily be described, and all roles involved can be trained in the "new rules." An implementation of supporting technology, though, might be an idea that belongs in the next category.
For this category, the next step will be to create the business cases that we will discuss in the next section.
3. Less defined ideas that **require more research** to see if they are feasible or require a rethinking of standards, application/risk landscape, or even a change in strategy. You definitely want to plan for a research phase that will then bring these ideas up to a level that allows the creation of a business case.
An example here can be to review the SLAs for the finance manager role in the context of their other tasks and might end in a redefinition of that role and hiring of additional resources.

This list will now be much easier to understand, and you will be able to articulate the benefits of each of the items to different degrees, which will determine the prioritization.

Develop business cases

While it might be easy to describe the benefits of the items in category 1, there might be more work to be done for the items in category 2, which is the near-/mid-term realization time frame.

The process for this is two-fold:

- Assessment of the idea when looking at the **levers of the process**. You might recall that we are asking the stakeholders what "a good process" will look like for them, and the idea is to get a quick assessment on the areas. For the examples in this segment, you will see that we have defined six levers, and the impact of the idea on these is indicated with a simple

traffic light visualization in the upper right-hand corner.
The result of this will determine if the idea will be taken off the list ("does not fit the good process profile") or determine the relative position on the list.

- For the upper 3–5 ideas on the list that require more calculations or discussion but fit the levers profile, you will have to spend some time calculating the cost and benefits. This will result in a decision if the result justifies the cost and effort. Sometimes new ideas might come up when the initial idea does not meet the threshold.

Let's take a look at three examples from our scenario.

Improve utilization (data modeling)

As mentioned above, the project was tasked to review the utilization of the financial manager role, but it was easy to do this in KNIME for all roles and to create the overview that is shown in figure 19.2 below.

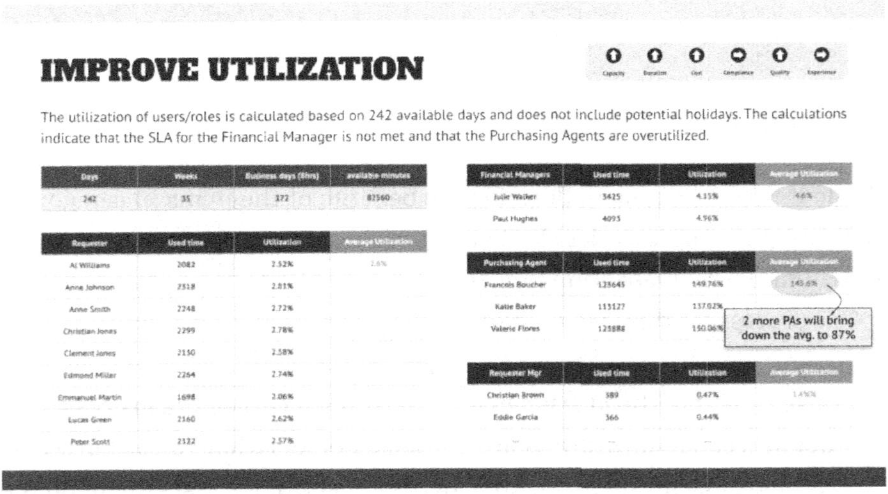

Fig. 19.2: Recommendation based on data model

As you can see, the SLA for the financial manager role was not met (4.6% instead of the desired 1%). It also indicated that the purchasing agent role is significantly over-utilized, with 145.6% on average.

Assuming that there will still be an overutilization of the resources after other ideas are implemented (training, supplier conversations, and retraining of users), the hiring of additional resources is needed. When looking at the levers, this will have a positive impact on capacity, time, and quality but will affect cost.

The exercise at hand is to calculate the needed number of additional resources (two new purchase agents will bring the utilization into a "normal" range of 87% for the individual and allow "extracurricular activities" such as vacation or training). Besides the salary for these roles, additional cost factors might be the cost of finding and selecting candidates, as well as their initial training and mentoring before they will be fully productive.

Supplier changes (process mining)

The next improvement idea is mostly based on the analysis in the process mining tool. By building dashboards for the different product areas, suppliers, and calculated case costs, it was clear that the higher-priced not only, well, were higher-priced than their competition but also created more disputes, which slowed down the process execution and drove up the case cost (and utilization of the two roles mentioned above).

This could be a symptom of complacency by the vendors ("that's how we have always done it"), a data problem, or simply unclear terms of what is expected from each player (a process problem). This requires more research, but when looking at the cost and time for the higher-priced suppliers compared to the lower-priced suppliers, there are significant savings in time and money to be expected by simply ordering as much from the lower-priced vendors as possible. Figure 19.3 shows the calculations for this scenario.

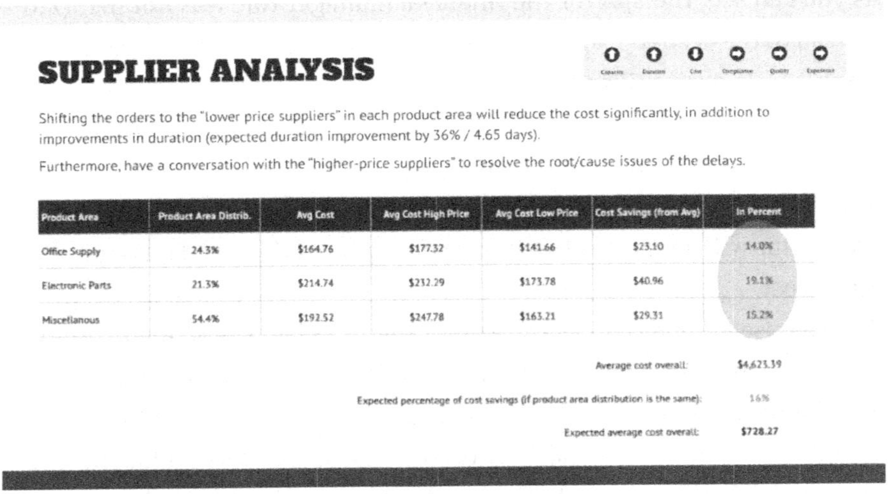

SUPPLIER ANALYSIS

Shifting the orders to the "lower price suppliers" in each product area will reduce the cost significantly, in addition to improvements in duration (expected duration improvement by 36% / 4.65 days).

Furthermore, have a conversation with the "higher-price suppliers" to resolve the root/cause issues of the delays.

Product Area	Product Area Distrib.	Avg Cost	Avg Cost High Price	Avg Cost Low Price	Cost Savings (from Avg)	In Percent
Office Supply	24.3%	$164.76	$177.32	$141.66	$23.10	14.0%
Electronic Parts	21.3%	$214.74	$232.29	$173.78	$40.96	19.1%
Miscellanous	54.4%	$192.52	$247.78	$163.21	$29.31	15.2%

Average cost overall: $4,623.39

Expected percentage of cost savings (if product area distribution is the same): 16%

Expected average cost overall: $728.27

Fig. 19.3: Recommendation based on process mining analysis

When looking at the levers of process for this scenario, there are no drawbacks, but only improvements, which makes this a very attractive improvement opportunity.

The long-term resolution to this issue might not be simple, though. However, reviewing the process with the suppliers and clarifying SLA expectations, combined with alerts and the change of the internal policy of whom to buy from, will bring some relief.

Going forward, the review of the supplier strategy and potential automation to reduce data errors (using AI agents to ensure that all information is not only available but also correct and reflecting the correct day-by-day prices for goods) requires more research and planning so that these sub-ideas might be separated on the list of improvement opportunities.

Process changes (process mining, modeling, simulation)

To determine the needed changes in the process, process mining alone will not give the answer.

The high-level assessment of the ideas is very positive. While there is some cost involved for automation, rollout, training, etc., the promise of reducing cost and time while improving capacity, quality, and compliance is enticing. Figure 19.4 shows the areas for process improvement.

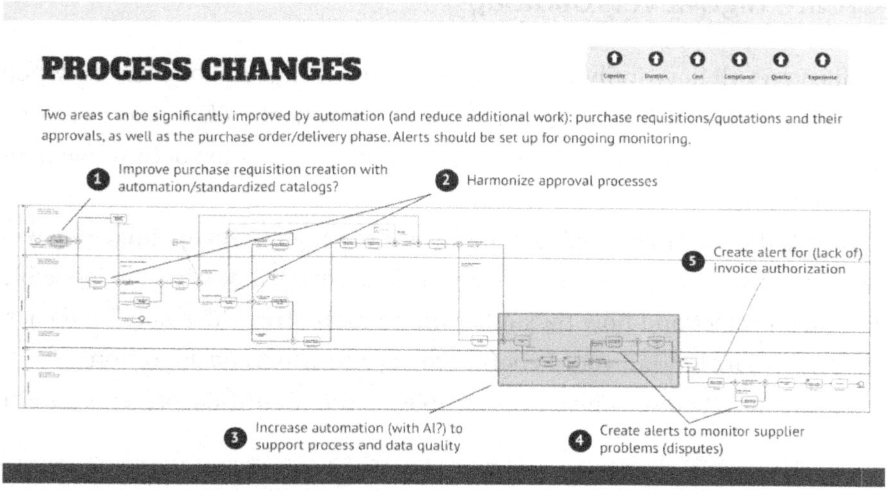

Fig. 19.4: Proposed process changes

Some process changes are quick and fast and already on the list (alerts and harmonization of the approval steps), but for automating steps that will inevitably come with changes in the process, as well as an impact on the application landscape or standards, there is more research needed.

As always, this should trigger the creation of a solution architecture that will start with depicting changes in the process and then the simulation of the future-state process(es). Depending on the way you design the changes, there might be different future-state scenarios with different costs associated with each (for the number of users, technical resource costs for machines or AI agents, and so on). Simulation is a good way to predict the KPIs that are relevant for the levers of the process, such as capacity or time.

My recommendation is to split the process changes into multiple ideas so that you can manage them separately on your list and determine the benefits for each independently ("worst case is that we have only this improvement" ... but that is better than throwing out a big idea completely and having no benefits realized).

Create high-level roadmap

While you will have individual conversations with your stakeholders when discussing the individual ideas, there is a low probability that everyone involved understands all the details of all items on your list of improvement opportunities.

Now is the time to "translate" the ideas into a high-level roadmap that will show all ideas as projects on a timeline and also determine who is responsible for each project and how they all relate to each other. The details of each project will include all your benefits, costs, and effort considerations.

Figure 19.5 below shows an example of the identified projects in our scenario.

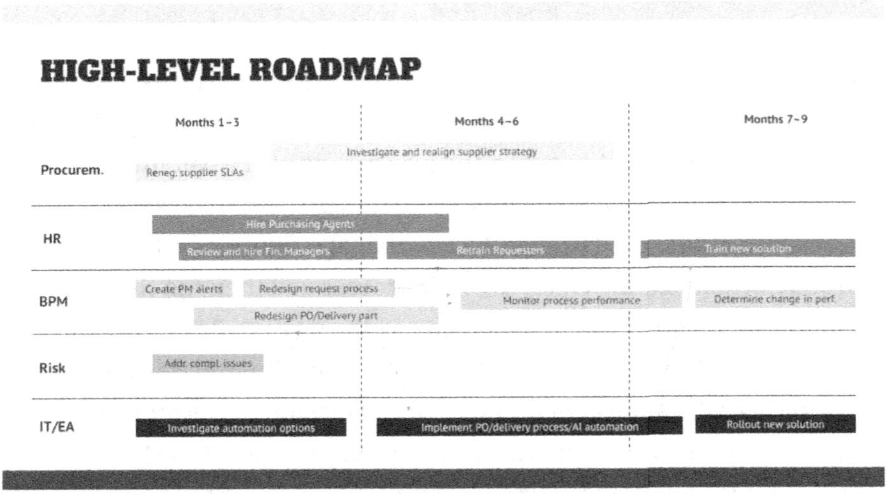

Fig. 19.5: Example high-level roadmap

For obvious reasons, you should socialize the high-level ideas of your roadmap with your stakeholders and work out with them where their priorities are and what impact it might have on your timeline and dependencies before it comes to the "big presentation" of your project results that culminates in the presentation of the roadmap shown above.

At that point, there should be no unknowns anymore that could lead to showstoppers. But the meeting with all different stakeholders involved has different purposes:

- First, you want to close out your process analysis project and get buy-in for your analysis results. The read-out presentation is the close-out of this project, and you want to capture requirements for future process analysis projects where you can apply your process mining skills.
  This part of the presentation should show what you've done, what the results were, and include answers to all questions and hypotheses. A "parking lot" list will capture all new requirements and ideas.
- Then you want to discuss the improvement projects on your roadmap. Please note that I say "projects" and not "ideas" or "opportunities" anymore. Only well-defined items go on the roadmap; everything else goes on the parking lot list.
  You will present the list of projects and give the background information about how you got to it (see the segment above in this chapter) and where they land on the prioritized list—reflecting the three categories from the top of this chapter.
- Once everyone understands each project and agrees on the benefits/cost/prioritization of each project, the planning phase begins. What you want is the buy-in from your stakeholders to your plan (as shown in figure 19.5). If there are discrepancies, for example in the timing of projects, then try to solve them directly in this meeting.
  In my experience, any delays that come from "offline conversation requests" will push out the plan by weeks or even months, and you open the opportunity for others to renegotiate their part of the plan, which will create even additional modifications.

- If you found an agreement, your main decision-maker will also be in the room and can give the approval for implementation directly. If that is not the case, I hope you have discussed the material with them before so that you just need to communicate the delta, and you don't have any further delays.

20

Process Mining in Real Life Summary

There you have it—a complete walkthrough of the six steps of successful process mining projects, based on the example of our fictitious medical device company.

You have noted that the steps build up onto each other and that you can do (and have to do) process analysis in more than just your process mining tool. However, the mining tools by themselves will bring you a good step forward and simplify data-driven process analysis immensely ... think of the "Digital Gemba," which is much simpler than the actual gemba.

And, as always, there are a few takeaways for this section as well:

1. The focus of this section was on setting up and applying the various analysis types. But, if you compare these with the six-step overview in chapter 4, you will notice that some steps outside the actual analysis steps were left out or not featured prominently. To avoid forgetting these, I created a Successful Process Mining Projects Checklist as a complimentary freebie to this book. Please refer to the next chapter ("Before you go...") on how you can access this and the other freebie from the book website.
2. As mentioned a couple of times throughout the book, it is critical that you develop a strategy for the process mining capability in your organization. You should plan for an organization-wide roll-out even if you just get

started with a single project, such as our Heartbeat example.

Getting the key objectives and decisions documented and developing a short- and long-term plan—even if you will iterate on it as needed—is beneficial and will simplify the communication to your stakeholders and "fast followers" immensely. One key example is the definition of an ideal process profile that will give direction to all your participants in this program, and you will be surprised how many opinions you will hear from the stakeholders when you bring them to the table.

3. We cut the second step of identifying and extracting the data from source systems short by design. There are so many integration technologies, and your IT landscape of source systems will differ from others, that it didn't make sense to go into details here.

   But you should also keep in mind that program speed is more important than technical beauty. You want to have short, 4-5 week projects and build momentum, experience, and excitement for this technology and approach. That is more critical than building a "professional solution" that works perfectly but delivers "meh" results.

   When doing short projects, you can always add more iterations to go into more details and find more or more defined opportunities. And moreover, when you design a new solution, you can always go back to the process mining project that delivered your as-is analysis and drill down into the aspects that you need further information about.

4. Data modeling is the key for process mining. Getting your data into shape and understanding it in detail is critical to every project (and you will spend quite some time on creating your data model). If your stakeholders cannot follow how you have assembled the data model, they will not trust what they see in your analysis.

   To accomplish this, use a visual data modeling tool that allows business people without coding experience to follow your explanations. In an ideal scenario, they will take on the task of creating a data model for their next project.

5. Process modeling and process simulation might be "old" approaches and be more human-centric, but they still have value. As you have seen

in chapter 16, by simply visualizing the process model and enhancing it with information from the data set, you can find initial improvement opportunities (such as the potential compliance issue in our example). And moreover, you can always apply the "Base Analysis" features of your modeling tool, as well as the simulation tool, to the process diagram to find potential problems. These tools also can visualize issues like bottlenecks nicely with data that will be created when you run these tools (like the visualization of bottlenecks in bar charts in Simulation).

6. The additional benefit is that you can also use the process model to not only show "how work should be done" in preparation for the conformance analysis in step 5, but you can (and should) use that diagram to create a future-state version of the process, "how work should be done in the future."
   That model can also be simulated, and the output can be used for the business case for the proposed change and the update to the risk & compliance documentation.
7. Process analysis in your mining tool is the second key aspect in your project. But you also have noticed that there is not one analysis type in process mining, but it is an assembly of analysis types. Which one to choose is the key question, which is driven by the hypotheses that you have discovered in step 1 together with your stakeholders. Generally speaking, you should have a high-level analysis (which will be similar in all projects in your organization and enable some comparison between the projects), a detailed analysis based on the hypothesis, and the additional analysis that you have to do outside your process mining tool as needed (such as a two-factor analysis in your data modeling tool in our example).
8. And, as with the process modeling and "traditional analysis" above, you might want to take a look at older process analysis concepts, such as Lean or Six Sigma, and adapt these to your analysis. Especially the concepts of "value" and "waste" will help you to focus on your analysis if your stakeholders have a hard time articulating hypotheses for (lack of) process performance.

The good news is that the stakeholders also gain more experience from the project, and it will be easier for them to articulate these when they go through the second or third iteration of the project.

9. Dashboards are your friend, but they are (currently) not in the shape they should be in process mining tools. What you want to look for is the story that you aim to communicate to your stakeholders, which might require a different dashboard than the one that you used for your initial analysis (one topic per dashboard). Then you would like to define a clear information hierarchy in your dashboard and also configure the widgets in a way that you use the best type for the message and also remove as much visual clutter as possible.
   The benefit of using dashboards is also that they will update going forward, and you will be able to demonstrate how your improvement opportunities, well, improved the process.
10. When it comes to deciding on your improvement opportunities, you should not see any surprises when you reach step 6, but you should have maintained a log of ideas and assumptions during the project (captured in Notion or any other collaboration tool that your organization might use). Within the project time frame, you might have so many conversations that you will forget the ideas if you don't write them down.
    This will lead to missing out on looking at the underlying data to confirm or reject them as part of step 5.
11. In addition to this, keep in mind that you need to show and explain two things to your stakeholders: the high-level impact of your suggestions (which might lead to the prioritization of your ideas) and the calculation of the ROI and benefits—quantitative and qualitative—as well as the efforts and costs for implementing the selected improvement opportunities.

IV

Before You Go...

21

Before You Go...

Another book done, and I hope I succeeded in bringing process mining (and process intelligence) closer to you. I also hope that you took the offer and used the example files and recreated the six steps in your tool of choice.

This book should have given you all the arguments and information needed to implement process mining as a useful addition to your process management discipline in your organization. However, there are also "freebies" coming with it—all the slides of this book as a PDF and a checklist for each of the six steps of the Successful Process Mining Projects approach. To download them, please head over to whatsyourbaseline.com/successful-process-mining-projects and enter the code **SUCCESSFULPROCESSMINING** once you click the freebies link at the bottom of the page. The QR code at the end of this chapter will bring you to that page.

What is next for process mining?

I am not an oracle and just have limited visibility into what the different vendors do, but I can see a couple of trends that won't go away and will impact organizations going forward:

- **Process mining will not go away.** Quite the opposite, once organizations have seen the usefulness of having a real-time performance dashboard, they will just grow their appetite for this information. It is just like having a speedometer in your car—nobody wants to give that up, except when you are in a racing car in a competition.

- **Process mining tools will change.** I can see two developments: on the one side, there is process mining technology already built into runtime tools (like ServiceNow), and it will disappear as the "monitoring component" of these tools. You won't even know that you are using process mining because all the connections and the analyses are predefined and ship with the product.
  On the other hand, process mining will be integrated into the larger BPM tool stack as vendors like ARIS, Signavio, or iGrafx already demonstrate today. And we can see this also happen at the market leader at the time of this writing (mid-2025)—they bought a modeling tool and market it as "Celonis Process Management" but still cannot support the full lifecycle as described in a previous chapter above.
- I also spoke with an analyst from one of the large analyst firms last year. He sees **some vendors destroying a separate process mining market by the need to have large and costly projects** that might or might not fulfill the expectations. The vendors need to make a lot of money to "feed the beast." The better way, however, is not to have one big project but to have a series of smaller, quick-turnaround projects. Instead of one 12-month project, you should have 12 one-month projects to demonstrate value quickly. Or, as one of my bosses once said, "Momentum is a deliverable." This will reduce the complexity (and thereby reduce project risk) and also create the necessary "buzz" and excitement in the organization. You can always do a second or third project with an even more precise data set and get more insights.
- **Process mining technology will be commoditized.** It will become "normal," and the cost will go down. This will have an impact on the pure players in the market because they won't make their needed margins anymore. Besides the two developments above (becoming part of a runtime tool and being integrated in a larger BPM tool stack), I can see academically oriented and/or open-source tools continuing as

standalone tools.

In my consulting work, I have already seen these developments—even though a large organization might have an expensive process mining tool, they are searching for a cheaper alternative "for project use" (aka small and quick turnarounds). This might open the door for new billing models that are less complex and more straightforward; for example, getting rid of the "cell phone bill" style of nickel-and-diming for everything and rather having a simple consumption billing model.

A separate process mining/process intelligence tool market, as Forrester has defined it today, will go away over time.

- But remember: **ideas are nothing without execution.** Data-driven analysis is here to stay, and it will grow even more over time with the help of AI tools. However, if you don't take action and implement the improvement opportunities, then all the analysis effort was for nothing. The better way is to build process mining into your solutions and then use it for permanent measurement, alerting, and benefits realization.

But there is more...

Process mining (and architecture in general) is a discipline that is not done alone. And by that, I mean that you should look out for other people in the discipline outside your organization and connect with them.

Here are a few resources that you might want to check out:

- The **What's Your Baseline? podcast**. Its mission is to demystify enterprise architecture and business process management (including mining, of course), and it is hosted by yours truly and J-M Erlendson. We publish on Mondays every other week and have a mix of conversations with guests about topics of the discipline, as well as things that are relevant if you work in or lead a practice. You can find it wherever you get your podcasts (Spotify, Google, Apple, Amazon, YouTube, etc.) or on our website at whatsyourbaseline.com/podcast.
- Also follow **What's Your Baseline on LinkedIn**, which is the main channel

to announce new episodes and other things. A lot of the conversations and connections to other practitioners start from here: https://www.linkedin.com/company/whatsyourbaseline.

- A good **aggregator of BPM-specific blogs and podcasts** was set up by Mirko Kloppenburg and can be found at https://bpm.today.
- And lastly, follow or **connect with me on LinkedIn** as well (https://www.linkedin.com/in/rolandwoldt) or **reach out via email** to roland@whatsyourbaseline.com.

I am looking forward to connecting and speaking with you!

whatsyourbaseline.com/successful-process-mining-projects

Example Files

In previous chapters, I mentioned the various tools needed to do the analysis yourself. For your convenience, here are the most important links again.

Download the tools and files

To follow the example that is used in Part 3, you need a few tools and a few files—all free of charge:

1. First you need to download and install your copy of **KNIME** from https://www.knime.com/downloads
2. Then you need to register for a free **ARIS Process Mining** instance at https://aris.com/process-mining/ (click the "Try it for free" button and go through the process). This will give you a limited instance of the tool, but you can follow the example except for the conformance check :-(
3. Finally, you can download the **source event logs and the initial process** from the accompanying web page here: whatsyourbaseline.com/heartbeat

Some additional information

Both tool vendors offer training for the tools that are listed below. It is pretty straightforward for ARIS Process Mining, but for KNIME you will have to select the correct role.

- ARIS Process Mining training can be found at https://academy.aris.com/

learn. When looking for training, watch out for the term "ARIS Process Mining" (maybe with an "SaaS" in the title), and don't look at the older version of the tool that was called "Process Performance Manager."

- The KNIME training can be found here: https://www.knime.com/learning/. At the point of this writing, there are four tracks offered—do the L1 and L2 courses for the "Data Analyst" track.

  The main website also contains getting started guides and cheat sheets in the "Resources" section. In addition to that, there are many videos about KNIME available on YouTube, and they also have very active communities/forums where you can find the right pointer to solve your problem (highly recommended).

Credits

Every book is an assembly of content from "somewhere." Here are the credits for the external material that I have used in this book.

- Title photo—the main photo of the title is by Kelly Sikkema on Unsplash; the collage is done by me.
- All tool screenshots are taken using the ARIS platform by ARIS, KNIME by KNIME AG, and Notion by Notion Labs.
- Icons in graphics are from The Noun Project.
- Additional input used in graphics in this book is linked in the corresponding end notes.

Praise for "Successful Process Mining Projects"

- "A must-read for forward-thinking organizations. This book demystifies the role of Process Mining—not just as a standalone technique, but as an integrated discipline within Business Process Management. With clear, practical insights, it empowers teams to drive real value and sustainable process improvement."
  Hanneke Loefs-Mos, Owner of OLCAMA Consultancy, EA/BPM Consultant[59]

- "Roland has an exceptional ability to convey complex technical topics in a clear, concise manner and with the right amount of pragmatic humor. His book 'Successful Process Mining Projects' offers readers a truly groundbreaking roadmap, covering everything from initiation and conception to the development of a robust business case, operational implementation, and sustainability.
  The abstract topic of Process Mining should be considered a strategic investment with measurable added value. This reference work allows you to do just that when communicating with C-level management and the operational level.
  Roland's contribution to the sustainable implementation of Process Mining is an excellent handbook and reference work that readers will enjoy picking up again and again for guidance and support during the introduction phase and beyond."
  Katrin Wetzel, Head of Business Process Management[60]

- "Congratulations on this new book. It's a book that gives knowledge and makes it easy to get started with implementing process mining on the first process.
  'Successful Process Mining Projects' highlights several things to avoid as well as what not to skip and guides you through a process to get success with process discovery of the as-is process. It also guides you through project planning, analysis techniques, input to business cases, and input for process improvement. The book also describes that it is not just one more reporting application, but it gives value and new insight in doing process mining analysis for the business."
  Per Bandsholm, Implementation Project Manager / Grundfos[61]

- "Roland's book is a very hands-on and easy-to-follow step-by-step guide for getting started with Process Mining.
  And from my personal experience, I cannot emphasize more that you really have to take care of and put some effort into the first step of defining the goal and strategy of your Process Mining project. If you are not setting the clear path and bringing in all the stakeholders and their understanding of the goals and why and how we do Process Mining, you will have to go back to the drawing board in the middle or the end of the project.
  That has happened to me once, and you will not want to do this. As always in projects, you have to put time and effort into the preparation so you will succeed and not end up like the 70% of the process improvement projects ending up in failure because you have no idea what to do."
  Martin Holling, Head of Process Management[62]

- "Roland's new book, 'Successful Process Mining Projects,' provides a great explanation of what Process Mining is and how you might benefit from it. This book builds on his previous book together with all his conversations with business process experts in the What's Your Baseline? podcasts.
  It gives excellent, real-world examples of how to evaluate, select, and

implement a Process Mining tool and how to set up a process mining center of excellence in an enterprise. I thoroughly enjoyed Roland's easygoing, sensible, pragmatic advice on how to avoid potential process mining pitfalls and how to succeed with Process Mining."
Tony Phillips, Business Architect[63]

- "Roland has done it again! Following his acclaimed book on architecture implementation, he now presents a practical guide for those interested in process mining/process intelligence.
  This engaging read is packed with valuable content, featuring a six-step approach designed to help you achieve tangible results from your process mining initiatives. The helpful project in Part III, 'Process Mining in Real Life,' allows you to apply your knowledge in a practical setting. Highly recommended!"
  Zbigniew Misiak, co-author of "Practical Business Process Modeling and Analysis," BPMTips.com[64]

- "Process mining projects aren't easy, but they're transformative when done right.
  'Successful Process Mining Projects' offers a clear, practical roadmap for getting started, grounded in real-world experience. Roland doesn't just explain the theory; he guides you through the full lifecycle of a project, from initial ambition and framing, through analysis and iteration, to delivering outcomes that matter.
  The book's strength is its structured, step-by-step approach, supported by a realistic worked example that readers can replicate using the provided materials. It's not just about running a one-off analysis; it's about embedding process mining sustainably, with the right questions, the right setup, and a repeatable framework.
  Whether you're new to process mining or looking to sharpen your delivery, this book will help you build projects that work."
  Nick Blackbourn, Head of Process Intelligence—UK[65]

- "Already with his first book on 'Successful Architecture Implementation,' Roland presented a complex subject in a user-friendly way. He succeeds in doing this again in his second book on Process Mining.
  Anyone who wants to introduce Process Mining in their company and who does not yet have in-depth knowledge about it should first acquire the necessary skills from Roland's book. Easy to read and yet covering all necessary areas in depth. I'm looking forward to his next book."
  Randolph Schütte, Product Manager Master Data[66]

- "Roland succeeds in demystifying process mining (and process intelligence) in this easy-to-read and follow book. He takes you on this enjoyable journey of improving your organization's processes with defined steps and walks you through how to implement mining into your program."
  Marc Mendelson, Business Strategy, Transformation & Customer Experience Executive; Big 4 Consulting Director[67]

- "With this book, Roland continues to share his experiences, which he gained in a career spanning 30+ years, and seamlessly connects to 'Successful Architecture Implementation.' This book equally serves people seeking the theoretical and academic understanding of Process Mining, as well as the practitioners that apply Roland's tested implementation methodology."
  Ulf Hennigsen, Enterprise Architect and Business Transformation Leader[68]

Notes

PREFACE

1 Gemba is the Japanese term for "actual place" and is used by Lean practitioners when they mean that they have to go to the place where value is actually produced and see it with their own eyes (a "gemba walk"). Go, look, see!

2 If you are interested in who these were, head over to https://www.woldt.de/project-portfolio and take a look.

3 And yes, it was not Celonis who invented that technology, but the "hot company in the BPM space before Celonis even existed." But as always, you need the right time to present your wares and do a good job in explaining the need for them.
So, kudos to Celonis for bringing awareness of this exciting technology to the mass market. Well done, and because of your efforts, more companies reap the benefits of it now!

ACKNOWLEDGMENTS

4 https://www.bpm360podcast.com/

5 whatsyourbaseline.com/podcast

WHAT IS PROCESS INTELLIGENCE (AND WHY YOU SHOULD CARE)?

6 Source: forrester.com/blogs/forresters-process-intelligence-landscape-preview-3-primary-trends-to-watch

PROCESS MINING IN THE SOLUTION/PROCESS LIFECYCLE

7 https://www.whatsyourbaseline.com/2019/09/the-value-of-enterprise-architecture-in-digital-disruption/

8 https://en.wikipedia.org/wiki/ITIL

9 https://www.apqc.org/

10 https://en.wikipedia.org/wiki/Supply_chain_operations_reference

11 I am sure that you were wondering, "When does he come to the point and talk about process mining?" and if this was a different book. Well, you are partially right—the majority of this chapter comes from my first book, 'Successful Architecture Implementation,' but process mining is an integral part of the lifecycle, and it is worth taking this larger detour.

If you are interested in 'Successful Architecture Implementation' and how the disciplines of BPM and EA (and their proper implementation) can help your organization, please visit whatsyourbaseline.com/successful-architecture-implementation.

THE SIX-STEP APPROACH TO PROCESS MINING

12 BPMN stands for "Business Process Model and Notation" and is an industry standard for depicting processes, as well as having a technical file format that allows data exchanges and automation. To learn more about BPMN, please see https://www.omg.org/bpmn/.

13 https://en.wikipedia.org/wiki/Responsibility_assignment_matrix

14 https://www.alteryx.com/

15 https://www.knime.com/

PROCESS MINING, TASK MINING, CASES VS. OBJECT-CENTRIC MINING ... OH, MY ...

16 These events are not limited to transactional systems that humans use but are also created when machines interact with them, such as AI agents or data from Internet of Things (IoT) setups.

17 You can download the specification of OCEL 2.0 here: https://ocel-standard.org/2.0/ocel20_specification.pdf

SELECTING A PROCESS MINING TOOL

18 https://www2.everestgrp.com/reportaction/EGR-2021-38-R-4377/Marketing

19 https://en.wikipedia.org/wiki/Gartner_hype_cycle

20 https://www.whatsyourbaseline.com/2021/11/episode-9-task-mining/

21 https://en.wikipedia.org/wiki/System_and_Organization_Controls

BUYING A PROCESS MINING TOOL

22 Some additional thoughts on this could include things like extra pricing up front, i.e., how many licenses are included and how much are extra licenses, the cost of the implementation fee, or an automatic yearly increase? You also might want to check on software updates—will these be included, or is that an extra fee as time goes on?

PROCESS MINING IMPLEMENTATION

23 Chapter 14 "Your Mission (Should You Choose To Accept It)"

24 You can find more information about "Successful Architecture Implementation" (and download a sample) here: whatsyourbaseline.com/sai.

25 http://apqc.org/pcf

26 Reference content is discussed in chapter 4 in more depth and can be an accelerator for your program.

GETTING STARTED SUMMARY

27 I just touched briefly on the different phases here, but I expanded this approach in a book of its own, which you can adapt to the smaller scope of a process mining implementation as well. For more information about my "Successful Architecture Implementation" book, please visit whatsyourbaseline.com/successful-architecture-implementation.

EXAMPLE SCENARIO

28 https://whatsyourbaseline.com/heartbeat

29 https://www.apqc.org/process-frameworks

YOUR MISSION (SHOULD YOU CHOOSE TO ACCEPT IT)

30 The notes that I took for my analysis of the example scenario can be found here: https://rolandwoldt.notion.site/Heartbeat-Process-Analysis-Questions-and-statistics-226ff6ba40e980deb3b8c1cd706bce91.

31 This graphic and all other slides are available as a freebie to this book at whatsyourbaseline.com/successful-process-mining-projects. Check the "Before You Go..." chapter for the password for the freebies.

32 This is not sponsored by any of the organizations mentioned. But if you are from one and have some money to spare, please reach out ;-)

33 https://www.knime.com/

34 For educational purposes, I will not use the component functionality in the examples used here. This might complicate a workflow look, but I think it is beneficial for the learning experience to show "how the sausage is made."

35 https://aris.com/process-mining/

36 As of writing this book in mid-2025 the e-learning material can be found here:
— https://academy.aris.com/learn
— https://www.knime.com/learning

37 whatsyourbaseline.com/heartbeat

DATA MODELING

38 A good overview of the data types can be found here: https://amplitude.com/blog/data-types.

39 https://en.wikipedia.org/wiki/ISO_8601

40 The example source and result files, as well as the KNIME workflow are available for download at whatsyourbaseline.com/heartbeat.

41 You find his book online and free of charge here: https://b-greve.gitbook.io/beginners-guide-to-clean-data.

42 In the "Process Mining, Task Mining, Cases vs. Object-Orientation ... oh, My ..." chapter.

43 And those little traffic lights in KNIME are very helpful to spot where the workflow breaks. Unfortunately, these things don't exist in ARIS Process Mining.

44 KNIME AG offers a proprietary server for this purpose called KNIME Business Hub: knime.com/knime-business-hub.

CREATING A REFERENCE PROCESS

45 I like to keep all modeling activities that happen in the repository of your tool stack (ARIS Connect in my case) in one chapter, even though these activities most likely happen in Steps 5 and 6 of our six-step approach.

46 I wrote a full chapter about importing existing data into an architecture tool in my first book, 'Successful Architecture Implementation.' You might want to check it out for more details.

47 There is also a dedicated chapter about base analysis in my 'Successful Architecture Implementation' book if you want to get more details about this technique.

48 https://www.leansixsigmadefinition.com/glossary/waste/

49 As you might have expected by now, I also wrote a dedicated chapter about simulation in my 'Successful Architecture Implementation' book :-)

50 Technically, you can run a simulation without any of these (well, maybe except the frequency), but the result might be pretty meaningless.

PROCESS ANALYSIS

51 https://en.wikipedia.org/wiki/Ishikawa_diagram

52 Another example of the search for a certain hypothesis confirmation can be done by building custom dashboards that are supported by additional visualizations in your data tool. The PDF of all the graphics (one of the freebies you can download) shows more examples than were used in this book, and they are labeled as "Extras" in the upper right-hand corner.

53 And yes, it sounds a bit like Minority Report.

54 The calculation is as follows: the time frame is 242 days/35 weeks. This gives us 172 working days (82,560 minutes). Julie worked 3,425 minutes (4.15% utilization), and Paul worked 4,093 minutes in the process (4.96% utilization). The average utilization, therefore, was 4.55%.

SIDEBAR: DASHBOARDS AND STORYTELLING

55 For more reading on this bad behavior and how to fix it, I recommend reading Edward Tufte's book "The Cognitive Style of PowerPoint" and Cliff Atkinson's book "Beyond Bullet Points" (which, ironically, is published via Microsoft Press).

56 Either in the form of a (Word) document or the shorter form of a slidedoc—the mix of document and slide deck that might be a more appropriate form of communication in our attention-deficit time. Learn more about slidedocs here: https://www.duarte.com/resources/books/slidedocs/.

57 Another printing term that has no applicable meaning anymore since all newspapers are digitally printed today. And even those will go away over the coming years ... another loss of culture, IMHO.

58 Source: https://www.powerusersoftwares.com/post/infographics-how-to-choose-the-best-chart-type-to-visualize-your-data.

PRAISE FOR "SUCCESSFUL PROCESS MINING PROJECTS"

59 https://www.linkedin.com/in/hanneke-loefs-mos-rc-22a860/

60 https://www.linkedin.com/in/katrin-wetzel-5a597513b/

61 https://www.linkedin.com/in/per-bandsholm-6322a66/

62 https://www.linkedin.com/in/martinhollingde/

63 https://www.linkedin.com/in/anthonynphillips/

64 https://bpmtips.com/ and https://www.linkedin.com/in/zbyszekmisiak/

65 https://www.linkedin.com/in/nickblackbourn/

66 https://www.linkedin.com/in/randolph-sch%C3%BCtte-9b16221b0/

67 https://www.linkedin.com/in/marcmendelson1/

68 https://www.linkedin.com/in/ulfhenningsen/

Made in United States
North Haven, CT
10 November 2026

22084521R00137